Plymouth South Middle School
Media Center

Macmillan
Revised
ENCYCLOPEDIA
of SCIENCE

Fuel and Power

Energy Sources, Engines, and Rockets

Clint Twist

Macmillan Reference USA
New York

Board of Advisors
Linda K. Baloun, B.S., Science Teacher, Highmore, SD

Merriley Borell, Ph.D., Consultant Contemporary Issues in Science and Medicine, Piedmont, CA

Maura C. Flannery, Ph.D., Associate Professor of Biology, St. John's University, Jamaica, NY

Owen Gingerich, Ph.D., Professor of Astronomy and History of Science, Harvard-Smithsonian Center for Astrophysics, Cambridge, MA

MaryLouise Kleman, M.A., Science Teacher, Baltimore County, MD

Irwin Tobias, Ph.D., Professor of Chemistry, Rutgers University, New Brunswick, NJ

Bibliographer
Sondra F. Nachbar, M.L.S., Librarian-in-Charge, The Bronx High School of Science, Bronx, NY

Advisory Series Editor
Robin Kerrod

Consultants
Dr. J. Beynon
David Brodie

Andromeda Staff
Caroline Sheldrick, Editor
Tamsin Osler, Assistant Editor
David West/Children's Book Design and
Ian Winton, Designers

Front Cover: Oil industry worker (*Sindo Farina/Science Photo Library*); Mine (*Tony Craddock/Science Photo Library*); Early light bulb (*Michael Freeman*)

Published by:
Macmillan Library Reference USA
A Division of Macmillan Publishing USA
1633 Broadway, New York, NY 10019

Copyright © 1991, 1997 Andromeda Oxford Limited
An Andromeda Book
Devised and produced by Andromeda Oxford Ltd,
11-15 The Vineyard, Abingdon, Oxfordshire, OX14 3PX England

Macmillan edition copyright © 1991, 1997
Macmillan Reference USA

All rights reserved. No part of this publication may be reproduced or utilized in any form or by any means, electronic or mechanical, including photocopying, recording, or by any information storage and retrieval system, without permission in writing from the publisher and copyright holder.

Library of Congress Cataloging-in-Publication Data

Macmillan encyclopedia of science. -- Rev. ed.
    p.   cm.
    Includes bibliographical references and index.
    Summary: An encyclopedia of science and technology, covering such areas as the Earth, astronomy, plants and animals, medicine, the environment, manufacturing, communication, and transportation.
  ISBN 0-02-864556-1 (set)
  1. Science--Encyclopedias, Juvenile. 2. Engineering--Encyclopedias, Juvenile. 3. Technology--Encyclopedias, Juvenile.
[1. Science--Encyclopedias. 2. Technology--Encyclopedias.]
I. Title: Encyclopedia of science.
0121 M27 1997
500--DC20   96-36597
                                                          CIP
                                                          AC

Volumes of the *Macmillan Encyclopedia of Science*
Set ISBN 0-02-8645561
 1 *Matter and Energy*   ISBN 002864557X
 2 *The Heavens*   ISBN 0028645588
 3 *The Earth*   ISBN 0028645596
 4 *Life on Earth*   ISBN 002864560X
 5 *Plants and Animals*   ISBN 0028645618
 6 *Body and Health*   ISBN 0028645626
 7 *The Environment*   ISBN 0028645634
 8 *Industry*   ISBN 0028645642
 9 *Fuel and Power*   ISBN 0028645650
10 *Transportation*   ISBN 0028645669
11 *Communication*   ISBN 0028645677
12 *Tools and Tomorrow*   ISBN 0028645685

Printed in the United States of America

Introduction

Energy and engines keep the wheels of civilization moving. This volume examines energy sources from fossil fuels to nuclear fission and the Sun. It describes the generation and use of electricity, and surveys power-producing machines from steam engines to rocket engines.

To learn about a specific topic, start by consulting the Index at the end of the book. You can find references throughout the encyclopedia to the topic by turning to the final Index, covering all 12 volumes, located in Volume 12.

If you come across an unfamiliar word while using this book, the Glossary may be of help. A list of key abbreviations can be found on page 87. If you want to learn more about the subjects covered in the book, the Further Reading section is a good place to begin.

Scientists tend to express measurements in units belonging to the "International System," which incorporates metric units. This encyclopedia accordingly uses metric units (with American equivalents also given in the main text). For detailed information on units of measurement, and on numbers, see pages 86–87.

Contents

INTRODUCTION	3
PART ONE: ENERGY RESOURCES	6
COAL	8
Fields and reserves • Coal formation • Surface mining • Underground mining	
OIL AND GAS	16
Fields and reserves • Formation • Production • Transportation • Refining	
NUCLEAR POWER	26
Safe energy? • How it works • Reactors	
SOLAR ENERGY	32
Sun power • At home • Solar electricity	
ENERGY FROM THE ELEMENTS	38
Nature's power • Waterpower • Wind power • Geothermal power • Tidal and wave power	

PART TWO: POWER PRODUCERS 46

POWER PLANTS 48
The generation game • Steam turbines • Nuclear power • Wind and water turbines • Powerhouses

USING ELECTRICITY 58
Transforming society • Distribution • Heating and lighting • Circuits and wiring • Motors

ENGINES 66
The age of steam • Steam engines • Gasoline engine • Engine systems • Other engines

JETS AND ROCKETS 76
The jet age • Jet engines • Rocketry • Liquids and solids • Rocket workhorses

UNITS OF MEASUREMENT 86

NUMBERS AND ABBREVIATIONS 87

GLOSSARY 88

INDEX 92

FURTHER READING 95

PICTURE CREDITS 96

Part One

Energy resources

Without fuels and other energy resources, human existence would be cold, dark, and primitive. The discovery of fuel, in the form of firewood, is as old as the discovery of fire itself; but there is far too little firewood in the world to support modern civilization. At present, most of our energy comes from burning fossil fuels: coal, oil, and gas. Coal is now our basic fuel; there is a plentiful supply and we have enough for hundreds of years. Oil and gas are much more useful, but we have very limited resources, enough perhaps for only another fifty to one hundred years.

In the 1970s, nuclear energy was widely believed to be the energy source of the future. Today, however, there is a large questionmark over the nuclear energy industry. Attention has now turned instead to renewable energy resources: solar energy, waterpower, wind power, wave power, tidal power, and our planet's own reserves of heat energy.

◀ Solar cells, also called photovoltaic cells, produce small amounts of electricity from the energy in sunlight.

Coal

Coal is a form of fossilized wood, and coalfields are the remains of great forests that existed hundreds of millions of years ago, during the Carboniferous period. For the last two hundred years, coal has been our most important industrial fuel, and today it supplies the major source of our electricity. In countries where large coalfields occur, coal mining is a leading industry, using some of the world's largest machines.

Coal is our most plentiful fuel, but total reserves will only last for a few hundred years. Although coal mining damages the environment, and burning coal causes pollution, it is likely to remain a vital source of energy for the foreseeable future.

SPOT FACTS

- The coal under the Earth's surface is the remains of plants that grew over a period of many, many millions of years.

- The United States has estimated coal reserves of at least 3 quadrillion metric tons.

- The Yallourn Mine in South Australia contains over 13 billion metric tons of brown coal.

- The United States is the world's largest exporter of coal; Europe is the main importer. Most coal, however, is used in the country where it is produced.

- The world's greatest resources of unmined coal are in the frozen wastes of Siberia.

- A seam of coal 1 m (3.3 ft.) thick was once, on average, a layer of peat 8 m (26 ft.) deep before it became buried and squashed hundreds of millions of years ago.

- A large bucket-wheel excavator can mine coal at a rate of 8,000 cu m (26,240 cu. ft.) per hour.

- About a quarter of the world's electricity is generated in power stations that burn coal.

- A big power station can burn 25,000 metric tons of coal every day.

- Coal contains sulfur. When the coal burns, the sulfur forms sulphur dioxide, which combines with water in the air to make sulfurous acid. Sulfurous acid is the main cause of acid rain.

- In South Africa, where there is plenty of coal but not much oil, gasoline is made from coal.

- If coal continues to be used at the present rate, the whole of this energy supply will be used up in just a few hundred years.

- In a gas explosion in a British mine in 1951, flames traveled some 12 km (over 7 mi.) along the underground tunnels.

- In 1966, 116 children died when coal from a waste tip slid down a hillside and buried their school, in Aberfan, South Wales.

- Underneath every coal seam is a layer of fossilized soil called "seat earth," in which trees once grew.

- Coal fires are very inefficient – most of the heat goes up the chimney.

- Heavy vehicles drive backward and forward over piles of coal stored at power stations, to squeeze out the air. Otherwise the coal would burn slowly.

- Before the age of steam, small coal mines or "gin-pits" were worked by just a few men and a donkey.

- When dinosaurs proliferated on Earth about 200 million years ago, during the Jurassic period, much of the coal that is mined today was already some 100 million years old.

- Between April 1993 and March 1994, a record 3 million metric tons of coal were mined from just one colliery in Yorkshire, in northern England.

Fields and reserves

Coal is a black or dark-colored mineral fuel that consists mainly of the element carbon. As well as carbon, coal contains a variety of hydrocarbons, often traces of sulfur, and moisture. Coal occurs as seams (layers in the rock strata) within the Earth's crust. The depth at which coal seams are found varies enormously. In some coalfields, the seam lies just below the surface. More often, coal seams are buried beneath hundreds of feet of rock.

The properties of coal have been known for at least 2,000 years, but coal was little used until the Industrial Revolution of the 1700s. Before then, firewood and charcoal provided virtually all the world's fuel needs. During the 1700s coal quickly replaced charcoal as an industrial fuel. In the 1800s, coal provided the driving energy behind the great Age of Steam. Today, coal is used mainly to produce electricity.

▼ World coal reserves are estimated at around 10 quadrillion metric tons. Most of them are in the Northern Hemisphere, although there are some important deposits in Australia. The United States, Russia, and China between them have over 60 percent of the Earth's total resources.

Energy from coal

Total world coal production stands at over 4 billion metric tons per year. Coal is still widely used in factory furnaces and domestic boilers, but most of the world's output is burned in power plants to produce electricity. Although it is sometimes considered an old-fashioned fuel, coal is in fact our biggest single source of electrical power.

World industrial energy resources

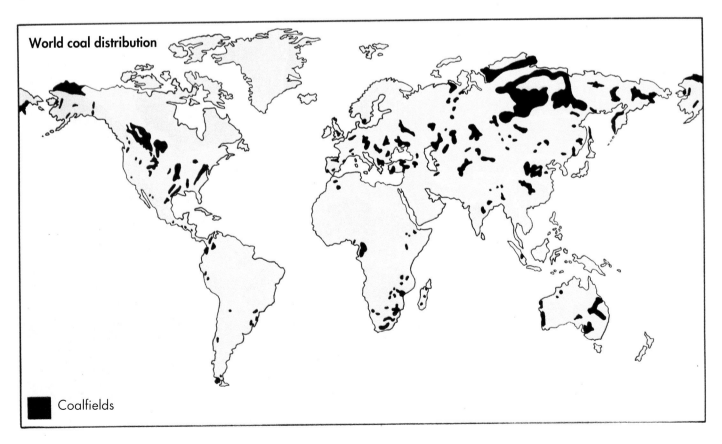

Coal formation

Coal was formed by the carbonization of trees and plants. When plants die, the carbon in their tissues is normally recycled back into the environment during decomposition. Carbonization occurs when dead plant material is subjected to heat and pressure over millions of years. The different grades of coal were formed by different combinations of time, heat, and pressure acting on the plant material.

Most of the world's coal was formed during the Carboniferous period, between 360 and 286 million years ago. During this time, large areas of the Earth's surface were covered with dense, swampy forests. Dead plants and trees that fell into the swamps did not decompose completely, but accumulated into thick layers of wet peat. When the swamps were later flooded by the sea, the peat became buried under layers of sediment. Over long periods of time, the sediment decayed further and slowly dried and hardened into brown coal or lignite. As further layers of sediment built up, increased heat and pressure caused lignite to turn into bituminous coal. In some instances, additional pressure turned bituminous coal into anthracite.

Anthracite coal is known as hard coal because of its rocklike appearance. Anthracite is the best-quality coal, and contains 86 to 98 percent carbon. It burns with a bright blue flame and gives off very little smoke, so is often used in built-up areas to help keep pollution down.

Bituminous coal is the most widely occurring grade of coal, and contains less than 86 percent carbon. Bituminous coal also contains the highest proportion of volatile material, which can be distilled into gas and coal tar. One of the most important discoveries of the Industrial Revolution was the process of baking bituminous coal in an oven to produce coke.

Lignite and brown coals are sometimes soft enough to be crumbled between the fingers. They contain less carbon than bituminous coal, and are generally compressed into pellets or briquettes before use.

▼ Fossil remains of the plant *Neuopteris fexuosa* found in coal. The plant lived in the Carboniferous period (360–286 million years ago). When the coal-forming material was laid down, the plant was trapped. Although the plant material decomposed, it left a clear impression in the coal seam.

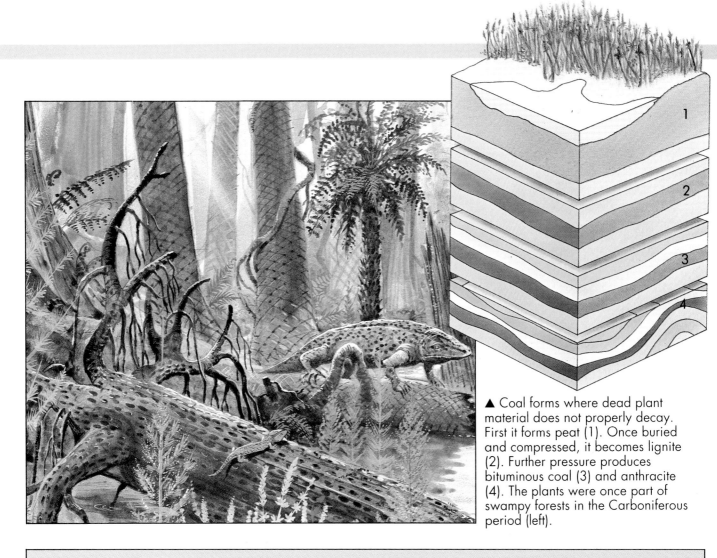

▲ Coal forms where dead plant material does not properly decay. First it forms peat (1). Once buried and compressed, it becomes lignite (2). Further pressure produces bituminous coal (3) and anthracite (4). The plants were once part of swampy forests in the Carboniferous period (left).

Nearly coal – peat

The water in swamps and bogs often does not contain enough oxygen or bacteria for normal decomposition to take place. Dead plant material slowly forms a layer of waterlogged peat. Over thousands of years, the layers build up and can reach 30 m (100 ft.) in thickness. Although peat is still some way from being coal, it can be used as a low-grade fuel. When it is freshly cut from the ground, peat is a black, slimy material containing about 70 percent water. After it has been dried, it is a crumbly, brown solid. When burned, peat gives off large quantities of thick smoke. In many rural parts of Europe, peat is still cut by hand in the traditional manner and is burned for domestic heating. In some countries it is cut by machines and used in small power plants. Most of Europe's peat reserves are now nearly depleted.

Surface mining

When a coal seam occurs just below ground level, it can be worked by surface, or strip, mining. The overburden – the rocks and soil that cover the seam – is removed, and the exposed coal is mined by mechanical excavators.

In the simplest surface, or open-pit, method of mining, the coal is dug out mechanically from an excavation. This method is used for relatively thick seams with little overburden. For thinner seams, a widely used technique involves cutting a series of trenches. When all the coal exposed by one trench has been removed, another trench is dug alongside and the overburden used to fill the first trench.

In Germany's Ruhr Valley, the coal seam lies very close to the surface, and it is only necessary to scrape aside the topsoil with draglines. But in parts of the United States, up to 60 m (200 ft.) of overburden lies above the coal seam.

In order to be economic, strip mining needs to be carried out on a very large scale. Some of the power shovels used can remove a vast amount of rock or coal – more than 150 cu m (200 cu. yd.) – with a single bite.

In general, it is the lower grades of coal that lie nearest the surface, and which are recovered by surface mining. This is especially true in the western United States and Eastern Europe, where huge deposits of brown coal and lignite are worked. Some high-quality coal is also obtained by strip mining, including almost half the anthracite produced in the United States.

The main problem with surface mining is the damage it creates to the environment, particularly if the coal lies beneath farmland. In many countries, strip mining is now carefully regulated. Some governments require that topsoil be removed and stored separately so that it may be replaced and the area replanted with a minimum of delay.

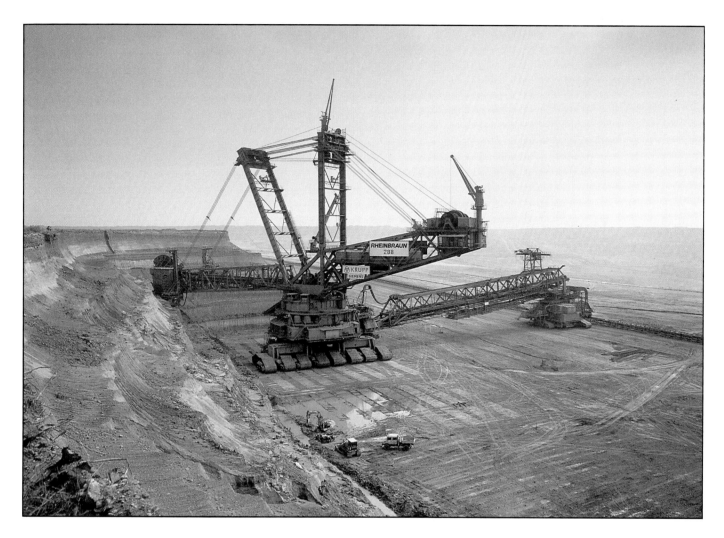

▲ A giant bucket-wheel excavator at a surface coal mine near Cologne, Germany. Operated by a crew of five, it can advance at a speed of 16 km (10 mi.) per minute, and is capable of moving 110,000 cu m (3.8 million cu. ft.) of coal, or earth, in a day. Its vast scale means that it has to be constructed on site.

◄ A strip mine in Australia, showing the extensive damage that surface mining can cause to a landscape.

▶ A power shovel in an American Strip mine. Even a very hard coal such as anthracite is soft enough to be worked by mechanical shovels and draglines. Brown coal and lignite are soft enough to be scooped out by bucket-wheel excavators.

Underground mining

▲ Typical hoisting gear at the pithead, often the only visible feature of an underground coal mine. This mine is in the Rhondda Valley in Wales. Early mines were not very deep, and coal was carried to the surface up a series of ladders.

In some countries underground mining is much more widespread than surface mining. Until relatively recently, the United Kingdom obtained the vast majority of its coal from deep under the ground, and mines were often more than 1,000 m (3,000 ft.) deep.

Underground mining is carried out along a series of horizontal tunnels and branching galleries. The exposed portion of the beam that is being worked is called the coal face. Some thin seams (less than a yard thick) are still worked by hand. Thicker seams are worked by modern coal-cutting machinery that can cut up to 6,000 metric tons of coal per day. Mechanical hoisting gear raises and lowers cages for miners and coal.

There are two basic underground coal-mining techniques: the room-and-pillar and the longwall methods. Room-and-pillar mining, which is widely used in the United States, involves removing coal from a series of underground rooms. Large pillars of coal are left in place to support the roof of the gallery.

Rotary coal cutter
Electric locomotive
Room-and-pillar mining
Hydraulic roof supports
Longwall coal face

Underground coal mine

▶ Modern mines have at least two shafts, one for coal, the other for miners and their equipment. A ventilation system pumps fresh air down one shaft and forces stale air out of the other. Miners may travel up to 2 km (over 1 mi.) by train to the coal face. Longwall mining enables coal to be cut by rotary cutters and be carried along conveyor belts into loading hoppers. At the surface, the coal is sorted and graded according to size and quality.

In Europe, the great majority of coal is obtained by the longwall method. Longwall mining involves digging two parallel tunnels into the coal seam. A gallery is then cut between them, and the roof is supported by hydraulic jacks. Coal is mined along the whole length of the longwall coal face. As the jacks are moved progressively forward, the roof of the gallery behind is allowed to collapse.

Water often has to be pumped out 24 hours a day to prevent flooding. Other major risks are underground explosions or fire, which can happen because of natural gas and coal dust suspended in the air. Great care is taken not to produce sparks, which could ignite the coal dust, and electrical equipment is heavily insulated.

Oil and gas

Crude oil, or petroleum, is one of the most valuable and versatile of Earth's buried treasures. Oil provides us with a number of different fuels, each of which is essential to modern civilization. Gasoline, aviation fuel, diesel fuel, and heating oil are all refined from crude oil. During the last 100 years, the search for oil has spread to ever more remote and difficult areas: hot deserts, offshore waters, and polar wastes.
More recently, natural gas has also emerged as an important fuel, and in many parts of the world it is piped directly into houses for domestic heating and cooking. Oil and gas are frequently found together.

SPOT FACTS

- Virtually all forms of powered transport now use fuels that can be extracted from crude oil.

- The Trans-Alaska Pipeline was designed to carry 2 million barrels of oil per day.

- More than 40,000 oil fields have so far been discovered. Fewer than 4,000 of them are of any commercial importance.

- The United States has already used up more than half its original oil reserves.

- Washing liquids, nylon, plastics, and fertilizer are all made out of oil.

- Crude oil is a mixture of substances (hydrocarbons) with molecules of different sizes.

- An airliner uses about 50,000 liters (52,834 qt.) of fuel to fly from Europe to America.

- To locate new sources of oil, geologists working for oil companies set off "mini-earthquakes." They explode some dynamite in holes in the ground, and then study the sound waves that bounce back upward from the layers of underground rock.

- Oil drillers pass a mixture of clay, chemicals, and water into new boreholes to keep the drilling equipment cool as it grinds its way down through the rock. They call the mixture "mud."

- More than 8,000 metric tons of crude oil are dumped into the sea each day.

- Some oil wells are 8 km (5 mi.) deep.

- Oil comes out of an oilwell because it is under pressure in the ground. As more and more oil comes out, the underground pressure falls. When the pressure falls too low, the oil stops flowing. Less than half of the oil in an oilfield can usually be extracted.

- 660 horses would have to work for 24 hours to provide the amount of energy that can be obtained from a metric ton of oil.

- The world's largest natural gas field is in Siberia. It lies only 1,200 m (4,000 ft.) below the surface.

- Natural gas (methane) can be transported by ship or by road tanker – but first it has to be cooled to -161°C (-288°F) to turn it into liquid.

- Gasoline, or petrol, is a mixture of about 300 different hydrocarbons – substances with molecules made of atoms of hydrogen and carbon.

- The gasoline used in the United States each year would fill 0.5 million swimming pools.

- Animal dung, human sewage, and waste plant material can all produce methane gas when they rot. It is called "biogas."

- In China, there are nearly 10 million biogas generators, providing methane from waste.

- Methane is a "greenhouse gas." Methane in the air contributes to global warming. In the last 30 years the amount of methane in the air has increased by some 50 percent.

Fields and reserves

By weight, oil and natural gas consist almost entirely of the elements carbon and hydrogen. Chemically, these two elements are combined into thousands of different compounds known as hydrocarbons. Oil and gas occur in natural underground reservoirs that may lie thousands of feet below the surface.

Both oil and gas have been known since ancient times, but they were very little used before the 1860s. The first oil well was drilled in Pennsylvania in 1859. Within a few decades, oil had also been discovered outside America. The most recently developed oil fields are in the polar regions of Alaska and Siberia.

Oil has become the world's most important fuel, which, when refined, is used for domestic and industrial heating and for powering the engines of our machines. Natural gas was once considered a waste product, but is now widely exploited as a heating fuel.

▼ Total world oil reserves are estimated at around one trillion barrels. The largest reserves are in the Middle East, which has 26 supergiant fields. Each supergiant field contains at least five billion barrels.

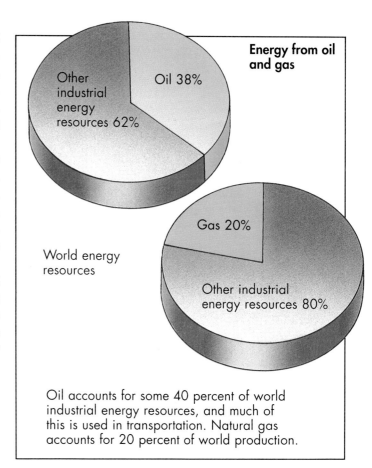

Oil accounts for some 40 percent of world industrial energy resources, and much of this is used in transportation. Natural gas accounts for 20 percent of world production.

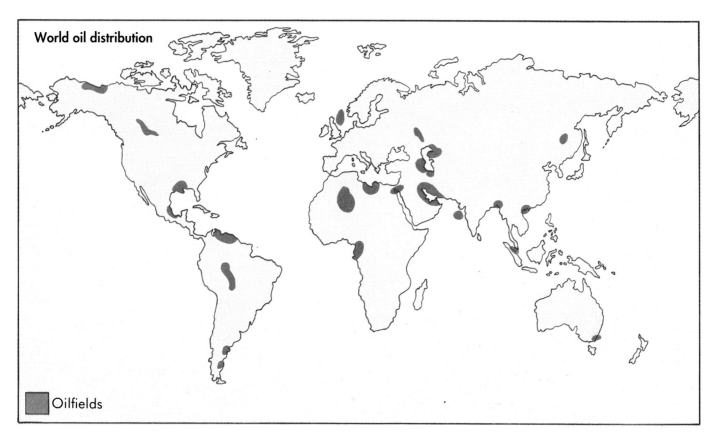

World oil distribution

Oilfields

Formation

Atoms of carbon and hydrogen join together in a number of characteristic arrangements to form hydrocarbon molecules. The smallest and simplest hydrocarbon, methane, has a molecule consisting of a single carbon atom surrounded by four hydrogen atoms. Larger and more complex hydrocarbons have molecules made up of a chain or ring of carbon atoms. Benzene is one of the most important ring compounds. Hexane consists of a chain of six carbon atoms.

▲ Pitch Lake in Trinidad, a naturally occurring deposit that has seeped out on to the surface, forming a lake of liquid bitumen (asphalt). Bitumen is one of the heaviest hydrocarbon compounds.

Oil was formed from the remains of tiny plants and animals that lived in the oceans many millions of years ago. The exact age of an oil deposit is hard to determine, because oil moves about easily and is not usually found in the rocks in which it was formed.

At some periods of Earth's geological history, the remains of algae and plankton accumulated on the seabed and were buried under layers of sediment. The sediment preserved the organic matter from the process of decomposition. Instead, it was transformed by the action of bacteria into a substance known as kerogen. Over long periods of time, further layers of sediment produced increased temperature and pressure, which "cooked" the kerogen and produced many different hydrocarbons. Depending on the exact recipe of the "cooking," crude oil can be thick and dark, or pale and thin.

Oil is chemically stable within the Earth's crust. But a number of factors cause it to seep through the rock and move from place to place. As a liquid, oil tends to move through the narrow spaces between particles of rock by capillary action. In rock that is saturated with water, movement is in an upward direction because oil is less dense than water. The oil often "floats" on top of a layer of water in the rock.

Like water, oil passes easily through permeable rocks such as sandstone and limestone, but cannot penetrate impermeable rocks such as slate. Some oil deposits migrate all the way to the surface, but most are eventually trapped by a layer of cap rock. The commonest type of oil trap is an anticline, an arch in the rock strata caused by the folding of the Earth's crust. About 80 percent of world oil production comes from anticline deposits.

▲◄ As it migrates upward through permeable rock, oil may become trapped by impermeable cap rock in a number of geological features: (1) an anticline, or dome; (2) a fault caused by Earth movements; (3) an unconformity (an abrupt change in the nature of the rock strata); (4) an intrusion (caused by a mass of rock pushing up from below); (5) a pinch-out (where the rock strata taper away); and (6) a salt dome (caused by the intrusion of a large pillar of rock salt).

Key
1 Anticline
2 Fault
3 Unconformity
4 Intrusion
5 Pinch-out
6 Salt dome

Natural gas is often found in association with oil, but also occurs as separate deposits. Most gas was formed by the same processes that created oil, although gas formation can take place over a wider range of geological conditions. Some gas has also been created by freshwater swamps or volcanic activity, and some has leaked from coal deposits. Gas is more mobile than oil, and may therefore migrate along different routes and be trapped in different geological formations.

The best-quality natural gas is "dry" gas, which consists of almost pure methane, with small amounts of ethane. Natural gas that also contains significant amounts of propane, butane, and pentane is known as "wet" gas.

Production

Modern surveying techniques can predict the location of an oil field with a fair degree of accuracy. The exact nature and value of the field has then to be proved by drilling a series of test holes to obtain samples. Only then can the location of the production wells be decided. Even on oil fields that are already in production, as many as four out of five wells that are sunk may turn out to be dry.

Under ideal conditions, natural pressure from gas and water trapped with the oil will cause it to flow up the well to the surface. Where there is no natural pressure, or where it has become exhausted, the oil must be pumped to the surface. In some cases, oil can be forced to flow up the well by pumping water into the oil field at a different location. Other techniques, such as underground explosions, have also been used to stimulate flow from unproductive wells.

About five percent of world oil production comes from heavy oil, which is much thicker than other crude oils. In many cases, heavy oil can be extracted only by pumping steam down the well to increase the rate of flow.

The deepest wells bring oil from more than 10,000 m (32,800 ft.) underground, but many fields lie within the first few thousand feet of rock. Faults and layers of drill-resistant rock are a much greater problem than depth. Such obstacles can be overcome by drilling at an angle, using a wedge-shaped attachment.

In some countries, oil is obtained from oil shale, a type of deposit in which oil is locked into the rock. Brazil, China, and countries of the former Soviet Union produce small quantities of "synthetic" crude oil from oil shale, but at present the process is expensive.

Undersea production

Extracting oil from beneath the seabed requires specialized marine technology to overcome the additional barrier of seawater. Offshore oil production began during the 1920s, but was restricted to simple platforms on stilts built in shallow coastal waters. During the 1970s, the search for oil was extended to the deeper waters of the North Sea and the Gulf of Mexico. Today, oil is produced from wells drilled in up to 300 m (1,000 ft.) of water.

◀ An oil industry worker in Argentina tightening a wrench. New technology in the industry has meant major advances in drilling technology in recent years.

▼ The pumps used to bring up oil are often known as nodding donkeys because of their ceaseless up-and-down motion.

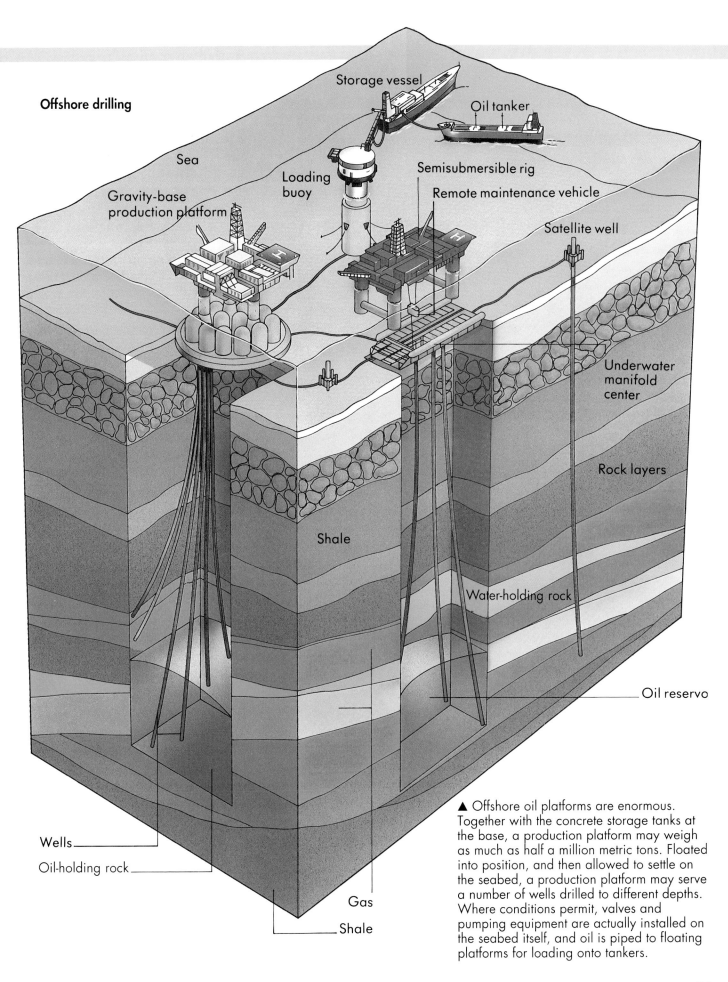

▲ Offshore oil platforms are enormous. Together with the concrete storage tanks at the base, a production platform may weigh as much as half a million metric tons. Floated into position, and then allowed to settle on the seabed, a production platform may serve a number of wells drilled to different depths. Where conditions permit, valves and pumping equipment are actually installed on the seabed itself, and oil is piped to floating platforms for loading onto tankers.

Transportation

Transporting oil

Oil is constantly on the move. Every day as much as 50 million "barrels" are transported somewhere around the globe. A "barrel" – the name of the basic unit used by the oil industry – is a volume of 159 liters (42 gallons). The scale of the global trade in oil demands transportation in bulk, either by pipeline or in ships that are specially designed for the purpose.

The largest ships, known as supertankers or VLCCs (Very Large Crude Carriers) measure over 300 m (1,000 ft.) in length, and can carry up to 3.5 million barrels of oil at a time. The oil is usually loaded and unloaded at specially constructed deep-water ports. In some places, for example the Persian Gulf, oil is first piped to floating platforms anchored some distance offshore. In some offshore oil fields, oil is loaded onto tankers directly from an underwater installation by a process known as wellheading. In other offshore fields, for example in the North Sea, the oil is carried from the production wells by pipelines to storage tanks onshore.

Generally, pipelines are used over short distances for local oil distribution (though there are some long-distance pipelines). The diameter of the pipes ranges from 20 to 120 cm, (8 to 48 in.) depending on the volume of oil to be carried. The construction of long-distance pipelines can present huge problems to engineers. But where these problems can be overcome, pipelines offer the cheapest form of bulk transportation.

One of the most demanding pipeline projects ever undertaken was the Trans-Alaska Pipeline. It was built to transport crude oil from Prudhoe Bay in the Arctic, south to the ice-free port of Valdez. During construction, engineers had to brave blizzards and temperatures as low as -50°C (-60°F). They had to cross countless rivers, deal with the problem of permafrost, or permanently frozen ground, and allow for possible earthquakes and migrating caribou.

▼ A liquefied petroleum gas tanker unloading its cargo into storage tanks. Natural gas can be stored in underground caves and disused mines.

Transporting gas

The best way to transport gas is to reduce its volume by cooling and compression. Natural gas can then be carried by high-pressure pipeline, or in liquid form on specially constructed ships. Liquefied natural gas (LNG), which is liquid methane, requires constant refrigeration. The liquefied petroleum gases (LPGs), such as butane and propane, liquefy fairly easily. They can be stored and transported at normal temperatures.

▼ (Main picture) A section of the 1,300-km (800-mi.) long Trans-Alaska Pipeline snaking across the frozen wastes of Alaska. In the northern part of the state, the pipeline is constructed above ground. This is necessary because the oil passing through is warm and the ground beneath is permanently frozen. Piping the oil through the ground would cause the permafrost to melt, leading to severe environmental damage.

▼ (Inset) A team of pipe-laying machines works in concert to lower a section of the Trans-Siberian Pipeline into a trench. This pipeline carries natural gas 6,000 km (nearly 4,000 mi.) from Siberia into Europe.

Refining

Crude oil is of little immediate use, because the individual hydrocarbons are completely mixed together. Useful fuels and other products have to be separated from the crude oil by refining. An oil refinery operates around the clock, processing up to 200,000 barrels of oil per day.

The main refining process is called distillation. Distillation makes use of the fact that different substances vaporize and condense at differing temperatures. Hydrocarbons with small, light molecules, for example methane and butane, are gases at room temperature. By contrast, the heaviest hydrocarbons, such as asphalt, are almost solid at room temperature. The liquid fuels – heating oil, diesel, gasoline, and kerosene, all vaporize at fairly low temperatures.

Refineries use a sophisticated process known as fractional distillation to separate crude oil into its component parts, or "fractions." The process is carried out inside a hollow steel tower known as a fractionating column. In large refineries these rise more than 45 m (150 ft.) above the ground, and contain 30 to 40 separate condensing plates.

Heavy oils like bitumen have big molecules, with long chains of carbon atoms. They are not useful as fuel for transport. But the heavy oils that are collected from a fractionating column can be "cracked" – their molecules can be broken into smaller ones. Heat and pressure can crack heavy oils to make lighter, more valuable hydrocarbons. Cracking happens more quickly using catalysts such as clays. This is sometimes called "cat cracking."

As well as fractional distillation, other techniques can separate some of the hydrocarbons from the rest of the crude oil. For example, hydrogen bubbling through crude oil carries away some of the lighter molecules, such as methane molecules, with it. Steam is used to remove these molecules from the hydrogen, so that the methane gas can be collected.

Solvents and acids can help to remove unwanted impurities from the oil fractions at the end of the refining process.

Refineries also produce petroleum spirits, waxes, greases, and carbon black for printing ink. Many other products are used as feedstock (raw material) for the petrochemical industry.

Fractionating column

◀ Crude oil is heated to about 350°C (660°F) before it is pumped into the column. Oil vapor then rises up the column through a series of steel trays, each of which contains a large number of condensation traps. The trays are maintained at slightly different temperatures, gradually getting cooler toward the top of the column. Different fuels and other products condense in the trays at different heights, and are tapped off. Any gas in the crude oil passes out of the top of the column, and is piped to a gas separation plant.

▶ The lights of an oil refinery glow against the night sky. An oil refinery involves several complex processes. After fractionating, some of the heavier oils are passed to the catalyst cracking plant to be broken into lighter grades. Sulfur may be removed from diesel fuel. The heavier fractions may be further processed, and a vacuum distillation plant produces lubricating oil and paraffin wax.

Nuclear power

Uranium is the rarest of all naturally occurring "fuels," and is also the most powerful. Inside a nuclear reactor, uranium can be made to release energy from the very heart of each atom. Most of this energy is in the form of useful heat. Nuclear power plants use only small amounts of uranium fuel, but produce large quantities of electricity.

Unfortunately, uranium is also our most potentially dangerous naturally occurring fuel. As well as heat, uranium emits potentially harmful radiation. Many people believe that nuclear power production presents a serious threat to the planet Earth. In some countries, the subject of nuclear energy is now highly controversial.

SPOT FACTS

- Just 5 metric tons of uranium fuel can provide the same amount of useful energy as 125,000 metric tons of coal or 500,000 barrels of oil.

- Uranium nuclei are the biggest nuclei found naturally on Earth.

- Uranium is radioactive. This means that its nuclei change, one by one, into smaller nuclei. The uranium changes into other substances. Ever since the Earth began, uranium nuclei have been doing this, but there are still plenty left.

- When nuclei in the atoms of a material absorb extra neutrons, the material can become radioactive. Fission in a nuclear reactor constantly produces neutrons. All the material of the reactor becomes radioactive.

- Nuclear power stations work by fission – splitting large nuclei. Some governments are trying to develop power stations that work by fusion – joining together small nuclei.

- Nuclear power stations generate electricity in the same way as in coal, gas, or oil-fired power stations. Energy is used to produce high-pressure steam that turns turbines.

- Unlike coal, gas, or oil-fired power stations, nuclear power stations do not directly produce carbon dioxide – the main cause of global warming.

- Highly radioactive nuclear waste must not leak into air or water for at least 10,000 years.

- Scientists studying the effect of radiation on the genes of plants have discovered that radiation makes the seeds of plants change, and this may cause them to form mutations.

- Nearly 90 percent of the radiation to which people are exposed comes from natural sources. Most of the rest comes from X-rays. Less than 1 percent comes from the nuclear industry.

- In 1986, there was an explosion at the Chernobyl nuclear power station in the Ukraine. For many years after the explosion, sheep in Scotland – thousands of miles away – were unfit to be eaten because tiny amounts of radioactive dust had fallen on the grass where they were grazing.

- After the Chernobyl explosion, firefighters and helicopter pilots worked to limit the damage as much as possible, knowing that the radiation could kill them. Many of them died.

- Thousands of people, all across Europe, may eventually die of cancer as a result of the cloud of radioactive waste that spread out from the Chernobyl nuclear power station.

- The "China Syndrome" theory refers to the possibility that, in the event of an accident in a nuclear power station, molten lead could burn through the center of the Earth and emerge on the other side.

- In France, a large number of nuclear power stations were built to a standard design in the 1970s and 1980s. Now France does not need to import so much oil or coal.

Safe energy?

The first experimental nuclear power plant started operating in the United States in 1951. Initially, there was considerable enthusiasm for this new source of energy. Nuclear power promised to supply the world with large quantities of clean, cheap electricity, especially in countries that lacked reserves of coal and oil. By the early 1960s, more than 100 nuclear power stations had been built. Developing nations were especially eager to acquire this new technology. Today, however, a question mark hangs over the nuclear power industry.

The great advantage of nuclear energy is that it uses very little fuel. The great disadvantage of nuclear energy is that it produces radioactive waste materials. In 1986, an accident caused an explosion at the Chernobyl nuclear power station in the Ukraine. The explosion created an enormous cloud of radioactive material that contaminated land and therefore food supplies over large areas of the former Soviet Union and Europe.

Nuclear energy

Nuclear power 5%

Other industrial energy resources 95%

World industrial energy resources

Nuclear energy accounts for up to 5 percent of world industrial energy resources. Most nuclear power stations are in industrially developed countries. France obtains 75 percent of its electricity from nuclear energy.

▼ The nuclear power station at Three Mile Island, Pennsylvania. In 1979, an accident caused the core of a reactor to overheat so much that it began to melt (inset). If the core had melted down completely, huge amounts of radioactive material could have been released, threatening the lives of thousands of people.

How it works

Nuclear energy is the energy released during the fission, or splitting, of the nuclei inside atoms of uranium or plutonium. Collisions with neutrons, which are particles from inside nuclei, can make some uranium nuclei split. The fission of a uranium atom also releases neutrons. Some of these strike other uranium atoms, causing them to split, thus releasing more energy and more neutrons. This process is known as a chain reaction. Uranium is the only naturally occurring element in which a chain reaction can take place.

Once started, a chain reaction keeps going until all the uranium is consumed. A nuclear reactor produces a slow, controlled chain reaction. The energy produced is released at useful levels over long periods of time. But a chain reaction can be made to spread through uranium almost instantaneously, producing the destructive power of an atomic bomb. Energy is released in huge amounts, but it cannot be put to any constructive purpose.

▲ Uranium fuel is loaded into a large number of cylindrical metal containers known as fuel rods. These are then packed closely together to form the core of the reactor. The shape of the core varies with different designs of reactor.

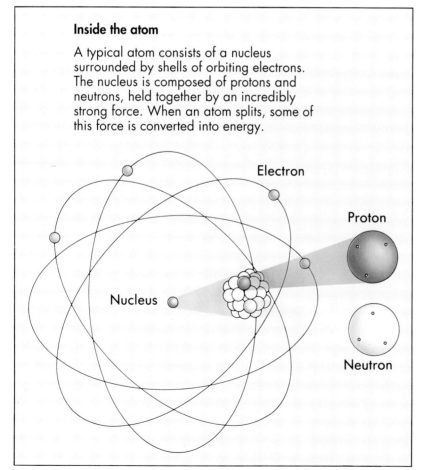

Inside the atom

A typical atom consists of a nucleus surrounded by shells of orbiting electrons. The nucleus is composed of protons and neutrons, held together by an incredibly strong force. When an atom splits, some of this force is converted into energy.

▲ The ore has to be refined in order to produce uranium fuel, which is shaped into thin rods or compressed into pellets.

▼ Mining uranium ore. Even good-quality ore may contain as little as 1–2 percent uranium. The largest deposits are in North America.

▲ A chain reaction starts with an atom of, say, uranium being struck by a stray neutron. Fission then produces more neutrons, which enable the chain reaction to proceed. The moderator helps control the speed of the reaction.

▲ The dangers of radioactivity mean that great care must be taken when transporting nuclear fuel. Special railroad cars are often used. These are designed to withstand a crash at speeds of up to 150km/h (90 mph).

The nuclear process

Using nuclear fuel requires a number of different processes. Uranium must be processed before it can be loaded into a reactor. After use, the fuel must be reprocessed for safe disposal of the high-level waste that is produced by fission.

▲ All forms of radioactive waste, including the clothing worn by workers, must be carefully sealed and stored. Waste with only a small level of radioactivity is often stored in steel drums. This is low-level waste.

▶ Highly radioactive waste must be treated with the utmost care. At this French installation, nuclear waste is being sealed into glass blocks. This reduces the risk of a leakage of radioactivity into the environment.

Reactors

A nuclear reactor has three basic components: a core, a coolant system, and a containment. The core produces heat, and the coolant system carries the heat away from the reactor. Most coolant systems operate under extremely high pressure, and the whole reactor has therefore to be encased in a strong reactor vessel. The containment is an outer covering, usually made of reinforced concrete, to absorb the radiation that spreads outward from the reactor core.

The core consists of fuel rods containing uranium, surrounded by a substance called the "moderator." Neutrons produced by fission in the fuel rods move around at random and at high speed. When they pass into the moderator material, they slow down. Slower neutrons are better than fast neutrons at causing more fission, and as a result, at keeping the chain reaction going when they travel back into the fuel rods. Graphite and water can both be used as moderator substances.

"Control rods" provide a way of adjusting the intensity of the chain reaction. They are made of material that is good at absorbing neutrons into the nuclei of its own atoms. Lowering the control rods deeper into the core soaks up more neutrons and slows down the chain reaction.

Enriched uranium

The majority of reactors currently in use run on uranium fuel that has been improved by a process of enrichment. Natural uranium is nearly all uranium-238, mixed with a much smaller amount of uranium-235. The basic

Pressurized Water Reactor (PWR) coolant systems

The coolant is the basis of a reactor's heat-transfer system. The coolant transfers heat from the core to a heat exchanger, where it is used to boil water for steam turbines. The pressurized water reactor operates at about 300°C (570°F) but the water remains liquid because it is kept under such high pressure.

Solar energy

The Sun represents a source of free energy. Many buildings already make some use of passive solar heating, and in many countries the Sun's energy is actively collected to provide hot water for household purposes.

The main drawback with solar energy is that it produces only low temperatures under natural conditions. In order to produce large quantities of electricity from solar energy, the heat energy in sunlight must be collected over a large area and concentrated at a single point. In countries with suitable climates, experimental solar-energy power stations are in operation.

SPOT FACTS

- The total energy of the sunlight reaching the Earth is equivalent to the output of about 200 million nuclear power plants.

- The first house to rely completely on solar heating panels was built in 1939 at an American university.

- Driven purely by solar power, the U.S.-designed vehicle *Sunraycer* achieved a speed of over 78 km/h (48 mph) in 1988. In the previous year it had won a 3,140 km (1,950 mi.) race for solar automobiles across Australia.

- The Sun works by nuclear fusion – joining together small nuclei of hydrogen to make bigger nuclei. In about 5 billion years' time the Sun will run out of hydrogen.

- Fusion of hydrogen nuclei happens only at very high temperatures. The high temperature of the Sun began when a huge cloud of gas collapsed into a ball. Ever since then, the temperature has been kept high by the energy released by fusion itself.

- The hydrogen bomb works by fusion, just as the Sun does.

- 10^{33} joules of thermonuclear energy radiate out from the Sun every year. Most of it spreads out into space, where it is used to power satellite equipment. Only one fifty-millionth of the Sun's energy hits planets like the Earth. If the Earth were to receive only a small amount more of the Sun's energy than it does now, the climate would be too hot to sustain most plant and animal life.

- Much of the energy that spreads out from the Sun is carried by X-rays, ultraviolet (UV) rays, and infrared radiation. These travel across space at the same speed as visible light. However, Earth's atmosphere acts as a barrier against these radiations, so very little reaches the Earth's surface.

- Plants make "carbohydrates" (such as sugar) using energy from sunlight. We, and other animals, get our energy from these carbohydrates. So all living things are in effect "solar powered."

- A third of the electricity used in Hawaii is generated by burning waste from sugar plants. The plants obtain their energy from the Sun.

- Houses in hot climates are often painted white to reflect the radiation from the Sun. Many have black solar panels on their roofs to absorb energy for heating water and powering electrical devices.

- The shiny black panels carried by satellites are solar cells. They use the Sun's energy to keep the satellite's electrical systems working.

- About 100 million solar-powered calculators are made every year.

- Some 90 percent of the houses in Cyprus have solar panels.

- A solar furnace in France focuses the Sun's light on to a small area to produce a temperature of 3,000°C (5,400°F).

- Solar cells are very cheap to run, but relatively expensive to buy.

difference between them is that uranium-238 has three more neutrons in its nuclei. However, uranium-235 is more useful in power stations because its nuclei can be made to split more easily by neutrons. Enriched fuel is uranium that has had some of the uranium-238 removed so as to increase the proportion of uranium-235.

Nuclear power stations tend to use water both as the moderator (to slow down neutrons) and as the coolant (transfering energy away from the reactor core to make steam, from which electricity is generated). The majority of water-cooled reactors use water at over 100 times the atmospheric pressure, and are known as pressurized water reactors (PWRs). Fast-breeder reactors operate at higher temperatures than other reactors, but have not proved commercially viable.

▼ The chart shows the electricity generating capacity of nuclear power plants ordered by European countries during the period 1956-85. Nuclear power became most popular when oil prices rose during the 1970s. Since the accidents at Three Mile Island in 1979 and Chernobyl in the Ukraine in 1986, nuclear energy has undergone a worldwide decline in popularity.

Hands off

Even small doses of radiation can be harmful to human health. Workers at nuclear power stations take every precaution to prevent their exposure to radiation. Inside the reactor building, workers wear heavy protective suits lined with radiation shielding. Delicate operations, such as removing depleted fuel rods, are usually carried out by remote control from behind heavily shielded walls.

At home

All houses and other buildings already make use of solar heating from sunlight shining on walls and rooftops, and through windows. Even in countries that have cool, cloudy climates, houses obtain as much as 20 percent of their space heating (room heating) requirements from the Sun. By incorporating passive solar technology, such as sloping windows and Trombe walls, this can be boosted up to 80 percent. This sort of heating is known as solar gain. Houses that are designed to make maximum use of solar gain are usually well insulated in order to keep heat loss to a minimum.

Producing domestic hot water from the Sun requires a solar collector. The commonest type is the flat-plate collector, which consists of tubes inside a sealed glass-topped box. The plate or tubing is black because dark-colored surfaces absorb heat better than lighter ones. The sealed box reduces heat loss. The water in the tubing may go directly to a hot-water storage tank. Alternatively, the water may flow in a closed system, and heat is transferred through a heat exchanger to the hot-water tank. This is more efficient than using the Sun-warmed water directly from the pipe.

Millions of these rooftop solar collectors are now in use worldwide, particularly in countries around the Mediterranean Sea.

▼ This British house makes good use of solar gain. Large sloping windows are the simplest form of solar technology. A sloping surface can receive up to 10 percent more solar energy than a vertical one. This principle is used in many solar-energy devices.

Sun power

Solar energy dwarfs all our other sources of energy. In less than one hour, the amount of energy that the Earth receives from the Sun is equivalent to the world's total energy output from other sources during an entire year.

Most of the Sun's energy is reflected back into space, but enough sunlight reaches the Earth's surface to make it worth exploiting as a source of power. On a summer's day in Michigan, for example, the energy falling on 1 sq m (about 1 sq. yd.) of sunlit ground is equivalent to ten 100-watt light bulbs.

Sunlight provides a constant source of energy for the Earth as a whole, but it is not evenly distributed over the planet's surface. Solar energy can only be exploited in bright light. Its greatest potential, therefore, is in hot countries that have clear skies for most of the year. But even in cold countries solar energy can be very usefully exploited.

The energy in sunlight can most easily be exploited in the form of direct heat. Rooms can be heated simply by letting the Sun shine in freely. Higher temperatures, needed to provide domestic hot water, require the active technology of solar collection panels.

Sunlight can also be converted directly into electricity using solar cells. A solar-powered aircraft was flown from the United Kingdom to France, and solar-powered automobiles have been built in several countries. Solar cells provide useful energy supplies to remote villages that are not connected to power stations.

▼ About 70 percent of the Sun's energy that reaches Earth is absorbed by the atmosphere, where it can be exploited as a source of power. Energy-saving designs that make use of solar power include solar panels to heat water, and solar cells to derive electricity from sunlight. In addition, electricity can be obtained from a windmill connected to a generator.

Solar energy

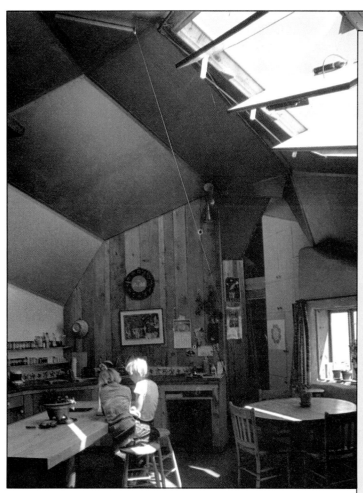

▲ This spacious house uses a number of light-sensitive switches to retain the heat obtained from solar gain. When the Sun is shining, the windows open and air can circulate through the house. When the Sun goes behind a cloud, the switches close the windows to trap the heat.

Evacuated-tube collectors can also be used to supply houses with hot water. An evacuated-tube collector may consist of a blackened glass cylinder inside a sealed glass tube. A vacuum inside the tube insulates the inner cylinder against heat loss. Heat is transferred by air or liquid circulating within the inner cylinder. A typical installation may contain 20 or 30 tubes connected together.

Evacuated-tube collectors are more expensive than flat-plate collectors, but can deliver about twice as much useful energy over a year. By surrounding the tubes with curved mirrors, or by focusing the Sun's rays through lenses, higher temperatures and greater efficiency can be achieved.

Active and passive

In "active" solar-heating systems, a working fluid, such as water, is pumped around a closed circuit. An example is a flat-plate collector where the Sun's energy is used to heat water and that heat is then transferred to a separate supply of water. A Trombe wall is a form of passive technology. Sunlight shining through the glass outer wall is absorbed by the dark-colored inner wall. Convection currents circulate warm air into the room, and draw cold air out.

Solar electricity

Producing useful quantities of electricity from solar energy requires a very large-scale installation. The most widely-adopted design is that of the solar field. This is already being used in several countries including Australia, Japan, Spain, Italy, and the United States.

A solar field consists of many rows of individual solar collectors. These are connected to a central heat exchanger that produces steam to drive an electric generator. The collectors are normally surrounded by curved reflectors, and are made even more efficient by the process of Sun tracking. Each collector is mounted so that it can be swiveled and tilted to be always facing directly at the Sun. Throughout the day, the collectors' positions are constantly adjusted by small computer-controlled motors.

The main disadvantage with the solar field is that heat energy is lost during the transfer from the collectors to the central heat exchanger. One solution to this problem is to concentrate the Sun's heat into a central collection point by using a circular field containing thousands of separate mirrors. The first central collection systems were experimental solar furnaces, which could reach temperatures above 3,000°C (5,400°F). During the 1980s, the first central collection power towers began operating. By this method, the Sun's rays are focused at the top of the tower, and heat is collected in a series of black-colored pipes containing liquid sodium. Heat exchangers at the base of the tower connect to the boilers, in which steam is produced to drive the generators.

Photovoltaic cells can be used to produce large quantities of electricity, but are expensive to make. Manufacturers are seeking ways to make them more cheaply. As prices come down, they will be used more and more.

▶ The world's largest power tower at Barstow, California. The tower itself stands about 90 m (295 ft.) tall, and the field of mirrors covers some 90,000 sq m (over 100,000 sq. yd.).

Solar cells

A solar cell, often called a photovoltaic (PV) cell, converts the energy in sunlight directly into electricity. An individual cell consists of two thin slices of silicon crystal sandwiched between two layers of metal. The top layer of metal is in the form of a grid so that sunlight can reach the upper side of the silicon. The two slices of silicon contain slightly different amounts of impurities, causing them to have different electrical states. Sunlight falling on the upper slice causes electrons to flow into it from the lower slice. This creates an electrical current that flows through the metal contacts. The photograph shows panels of solar cells mounted on a research satellite. In space, the cells will receive the full strength of the Sun's rays, and will be able to operate at maximum efficiency. On the Earth's surface, however, even strong sunlight has had most of its energy filtered from it by the atmosphere.

Energy from the elements

Our planet is rich in natural sources of energy. Waterpower, wind power, and wave power provide further opportunities to use the energy of the Sun. Solar heat powers the water cycle, which provides rainfall and running water. Uneven heating of the Earth's surface causes the winds to blow, and at sea the wind creates waves. The daily rise and fall of the tides, however, is caused by the gravitational pull of the Moon.

The motion of water, wind, waves, and tides can all be harnessed by machines, with varying degrees of success.

Geothermal energy is the heat energy found in rocks deep below our planet's surface. This source of energy can be used more directly.

SPOT FACTS

- The first use of wind turbines to generate electricity was in Denmark in the 1890s.

- The world's first power plant to produce electricity from geothermal energy was built in 1904 at Larderello, Italy.

- About a fifth of the world's electricity is generated in hydroelectric power stations.

- There are nearly 20,000 large and small hydroelectric schemes in the United States.

- Hydroelectric power stations usually rely on water trapped in huge reservoirs. The weight of water in these reservoirs, acting on the rock below, can be enough to cause earthquakes.

- To make room for the Three Gorges Dam in China, several million people had to leave their homes.

- Waterpower is the most widely used of the natural energy sources, but only about one-third of suitable sites have so far been used.

- In the Earth's crust, the temperature rises by about 1°C (33.8°F) for every 30 m (98 ft.) of depth. This heat is generated by natural radioactivity.

- The world's most successful tidal power station is that on the Rance River, in Brittany, France. It began operating in 1966.

- In places near volcanoes, such as Iceland and New Zealand, hot rock lies close to the surface of the Earth. These hot rocks are used to generate "geothermal energy."

- Most of the homes in Iceland are heated by water that rises to the surface from hot rocks that lie deep underground.

- Tides are caused by the pull of the Moon's gravity on the Earth's seas and oceans.

- Tidal power stations exploit the movement of water caused by the ebb and flow of the tide. Turbines are placed in barriers across river estuaries. The turbines have to be able to work in both directions – as the water flows in and as it flows out again.

- Sailing boats and wind turbines use energy from the wind. Though fossil fuels will one day be used up, winds will blow for as long as the Sun shines on the Earth.

- There are some 4,000 wind turbines on the wind farm at Altamont Pass, California. These wind turbines are capable of producing about 20 percent of the electricity that a conventional power station could generate.

- When nuclear power stations produce more energy than people need, the spare energy can be used to pump water into high reservoirs. Then when the need for electricity increases, the water is used to power hydroelectric generators.

Nature's power

Nearly all of nature's power that we harness comes in the form of movement. It is largely a case of converting one sort of movement into another. The movement of running water and the rushing wind is converted into rotation by machines based on the wheel.

A waterwheel is placed edge-on into a flow of water. The wheel is turned by the force of the flow against blades set across the wheel's rim.

In a traditional windmill, the wheel takes the form of a number of angled blades, or sails, that face the wind. A windmill turns because the wind is deflected by the angled blades as it flows through the wheel.

Waterwheels and windmills have been in widespread use for at least 2,000 years. Since Roman times they have provided useful energy for milling grains for flour, or for pumping water for irrigation and drainage. At the beginning of the Industrial Revolution, water power provided most of the energy that drove the spinning wheels and other machinery in the first factories.

The twentieth century saw a further development in the exploitation of wind and water. Water power and wind power are now used to drive turbines, which in turn are used to generate electricity.

▶ Traditional windmills provided a steady source of low-speed rotation that was very useful for certain tasks such as grinding grain or pumping water.

▼ An ancient waterwheel in Syria, possibly dating from Roman times. Waterwheels could be almost any size, but the materials used were generally not very strong.

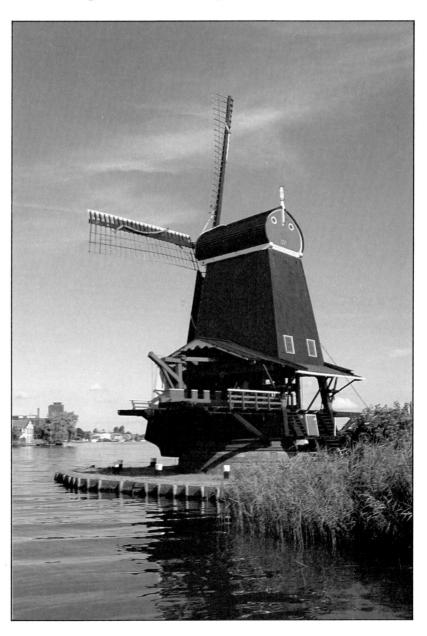

Waterpower

Water is the most useful source of natural power because it is the easiest to control. Running water can be channeled along sluices and through pipes. More importantly, a river can be blocked by a dam, creating a reservoir that can store huge quantities of water. Water from the reservoir can then be released as and when it is required.

Waterpower is harnessed to generate electricity in hydroelectric power plants. These are usually situated at the base of a large dam. The best locations are the narrow, steep-sided river valleys found in mountainous areas. A dam across such a valley can create a deep reservoir. Large-scale projects may involve more than a simple dam and reservoir. In the Snowy Mountains of Australia, the waters of the Snowy River have been diverted by a series of underground tunnels to some 16 hydroelectric plants.

Waterpower can also be used to store surplus energy from other power plants. This is done in pumped-storage plants, which use two separate reservoirs at different levels.

During normal operation, water from the upper reservoir is used to drive turbines to produce electricity. After passing through the turbines, the water is stored in the lower reservoir. Whenever there is a surplus of electricity from other power stations, it is used to pump water from the lower reservoir back up into the higher one. Demand for electricity is at its highest during the day. This means that, in most stations, pumping is done at night, and generating during the day.

▼ Construction workers inside the tunnels at a hydroelectric plant. The worker on the right is in the main water-supply tunnel. The worker on the left is standing at the mouth of an intake pipe leading to a turbine.

Hydroelectricity

Waterpower contributes about 7 percent of world industrial energy resources, all of it in the form of hydroelectric power. Large hydroelectric plants supply about 20 percent of the world's electricity. Compared with waterpower, energy from other natural sources does not significantly contribute to world output.

▼ Dams are normally curved, as this gives them greater strength. The level of water in the dam is carefully monitored, and excess water is diverted down the spillway. The main water intakes, which are called the penstocks, are usually located about midway down the dam. In this illustration, the turbines are mounted with a vertical axis, and each drives a separate generator, which is located directly above the turbine.

Hydroelectric power plant

Wind power

Unlike water, the wind cannot be controlled or stored. Wind power must be exploited where and when it occurs naturally. Until recently, the wind was used mainly to drive small pumps for agricultural purposes.

The amount of power produced by the wind increases as the cube of its velocity. This means that a doubling of the wind speed produces eight times as much power. In general, wind speeds increase with altitude. At 10 m (30 ft.) above the Earth's surface, the wind speed is about 20 percent greater than at ground level. At 60 m (200 ft.) up, it may be 50 percent greater.

Traditional windmills were designed for use with fairly low wind speeds. The materials they were made from (wood and cloth) were not strong enough to withstand high winds.

Modern windmills, which are usually called wind turbines, are designed to operate at much higher velocities. As a result, they produce far more power, and can be used to generate electricity. There are two main types of wind turbine. The horizontal-axis turbine has the same basic layout as a traditional windmill. Instead of sails, it has a rotor shaped like an airplane propeller.

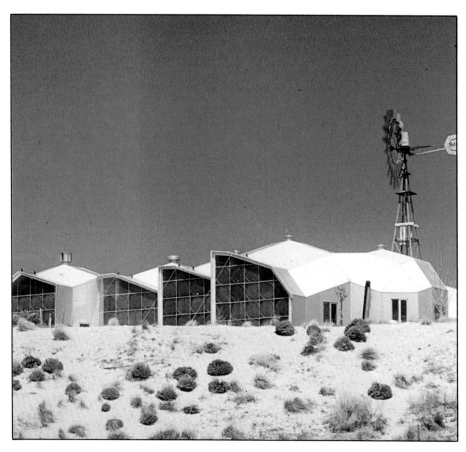

▼ Small wind turbines are used throughout the world to pump water and generate small amounts of electricity, particularly on farms. The commonest design uses a rotor consisting of a large number of metal vanes. The rotor turns on a horizontal axis. The whole of the turbine assembly is on a swivel mounting so that it can be turned to face the wind by the rudder.

▶ A wind farm in California, consisting of many rows of small Darreius wind turbines. Darreius turbines are easily recognizable by their distinctive shape, and can operate in wind coming from any direction. This particular wind farm is situated in a high mountain pass, where the winds are unusually strong and steady. Careful siting is the key to wind farming.

A vertical-axis wind turbine rotates on a shaft that is vertical to the ground. It normally has only two blades, mounted vertically at each end of a horizontal rotor.

The Darreius wind turbine is an advanced vertical-axis design that uses two curved blades. There is no separate rotor, and the blades are attached at each end of the shaft.

Wind turbines of both basic types are now in operation in many countries throughout the world. The largest ones are more than 100 m (over 300 ft.) tall. At full speed, the tips of the blades travel at up to 400 km/h (250 mph).

Wind turbines are positioned where winds are strongest, often on hills and cliff tops. On a wind farm, a large number of wind turbines, up to 30 m (100 ft.) tall, are placed near each other, each containing a separate electrical generator. Alternatively, just one or two large turbines may be placed at each location. Taking advantage of wind power may mean building wind farms offshore, where wind speeds are generally higher than over land. However, wind turbines are large and often noisy. Many people believe that they spoil the peace and scenery of the countryside.

Geothermal power

The Earth's crust is the thin, solid outer layer of our planet. Below about 100 km (60 mi.), heat from natural radioactivity is sufficient to keep rock in a softer state. Near the surface, the temperature rises about 30°C (over 50°F) for every 1 km (0.6 mi.) of depth of the ground. In areas of volcanic activity, this can increase to 80°C (nearly 150°F) per 1 km, and higher temperatures occur much closer to the surface.

In a few rare instances, in the United States, Japan, and Italy, this heat boils underground water, which rises to the surface as dry steam. This steam can be trapped and used to drive turbines. In California, the Geysers power plant has been built on top of a vast underground reservoir of dry steam. When operating at full capacity, the Geysers power plant supplies nearly as much electricity as is required by the city of San Francisco.

In most cases, superheated water remains trapped underground. When brought to the surface by wells, the water boils and the steam is used by turbines. Several countries, as far apart as Mexico and the Philippines, already produce electricity in this way.

Low-temperature geothermal heat has been used for thousands of years in public baths and health spas. In some parts of the world, hot water from volcanic springs and geysers is now a major source of domestic heating. In Iceland, more than two-thirds of the population now heat their homes with natural hot water. Other countries that make use of geothermal heating include the United States, Russia, China, Japan, France, and Hungary.

Even when there is no naturally occurring underground water of suitable temperature, geothermal energy can still be harnessed. Hot, dry rocks may be used soon in many countries as a huge underground boiler.

▼ (Below right) The Geysers geothermal power plant in California supplies electricity to a city of half a million people. (Below left) This design is for a power plant producing electricity from hot, dry rocks. Two wells are drilled some distance apart, one deeper than the other. The surrounding rock is then fractured with explosives to produce a large number of heat transfer surfaces. Cold water is pumped down the deeper well into the fissured rock, where it boils. Steam is tapped off by means of the other well.

Tidal and wave power

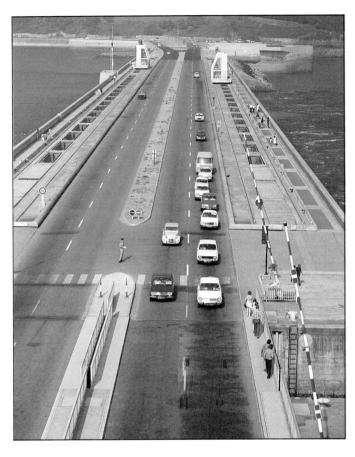

◀ A tidal-power dam across the estuary of the Rance River in France. At 750 m (820 yds.) long, the dam forms the world's largest tidal-power plant, with 24 turbines, which can operate when the tide is flowing in either direction.

▼ Wave energy converters: (1) The Salter "duck" uses a string of floating ducks hinged on a shaft; a full-sized duck would measure 25 m (82 ft.) across. (2) A development of the air-bag principle. The motion of the wave pumps air from one side of the converter to the other. (3) Each wave-contouring raft would cover at least 5,000 sq m (6,000 sq. yd.).

1 Salter duck

2 Air-bag device

3 Wave-contouring raft

When the motion of the tides is channeled by a natural feature such as a river estuary, it produces a very strong flow of water. The power of the tides can then be harnessed by building a dam equipped with turbines across the tidal flow. There are tidal-power dams in France, Canada, the former Soviet Union, and China.

The ceaseless motion of ocean waves also represents a potentially valuable source of energy. A number of wave-power devices have been invented and tested as models, but none operate yet as commercial power stations.

There are two basic approaches to wave power. The simplest designs place a line of wave-energy converters across the path of incoming waves. The energy converters float on the surface and consist of two sections hinged together. The motion of the waves operates the hinge, which powers pumps that drive turbines.

Other designs place the energy converter edge-on to the waves. The converters contain a number of air bags. As the wave travels along the converter, it squeezes the bags and forces air through turbines.

Tidal and wave power

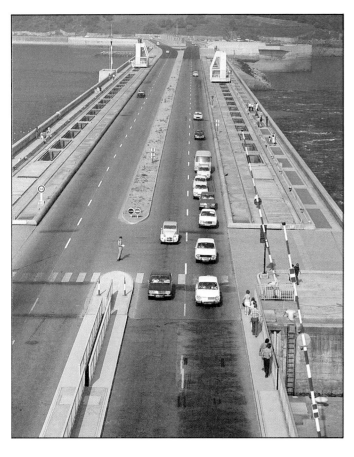

◀ A tidal-power dam across the estuary of the Rance River in France. At 750 m (820 yds.) long, the dam forms the world's largest tidal-power plant, with 24 turbines, which can operate when the tide is flowing in either direction.

▼ Wave energy converters: (1) The Salter "duck" uses a string of floating ducks hinged on a shaft; a full-sized duck would measure 25 m (82 ft.) across. (2) A development of the air-bag principle. The motion of the wave pumps air from one side of the converter to the other. (3) Each wave-contouring raft would cover at least 5,000 sq m (6,000 sq. yd.).

1 Salter duck

2 Air-bag device

3 Wave-contouring raft

When the motion of the tides is channeled by a natural feature such as a river estuary, it produces a very strong flow of water. The power of the tides can then be harnessed by building a dam equipped with turbines across the tidal flow. There are tidal-power dams in France, Canada, the former Soviet Union, and China.

The ceaseless motion of ocean waves also represents a potentially valuable source of energy. A number of wave-power devices have been invented and tested as models, but none operate yet as commercial power stations.

There are two basic approaches to wave power. The simplest designs place a line of wave-energy converters across the path of incoming waves. The energy converters float on the surface and consist of two sections hinged together. The motion of the waves operates the hinge, which powers pumps that drive turbines.

Other designs place the energy converter edge-on to the waves. The converters contain a number of air bags. As the wave travels along the converter, it squeezes the bags and forces air through turbines.

Part Two

Power Producers

Within the last 100 years or so, the supply of electricity has spread all over the world. As well as providing heat and light, electricity powers hundreds of different kinds of industrial machines and home appliances. For most people, this electricity comes from huge power stations that transfer energy from fuel or moving water into the rapid spin of turbines and generators.

Fuel is not just used to generate electricity. It is used for all kinds of useful work, such as powering vehicles and pumping water. The first useful fuel-burning engines were steam engines, but they have been replaced by internal-combustion engines. These use the rapid burning of fuel inside cylinders to move pistons backward and forward. Jet engines and rocket engines manage without pistons at all – using expansion of burning fuel to push directly on the plane or the rocket itself.

◄ Porcelain insulators on a pylon carrying high-tension electricity transmission lines in the Snowy Mountains, Australia. The electricity is produced by a hydroelectric power plant.

Power plants

Electricity is produced by the constant high-speed rotation of turbine-driven generators. The story of electricity is also a story of turbines. The widespread use of electricity became possible only with the invention of the steam turbine in the late 1800s.

Today, power plants equipped with steam turbines generate about four-fifths of the world's electricity. The remainder is produced by water turbines in hydroelectric power plants. During recent years, there has been considerable interest in wind turbines as a source of electricity. This kind of power generation is less harmful to the environment. In the United States in particular, thousands of wind turbines have been installed. However, the cost of wind power remains high, and so far it makes only a small contribution to world energy production.

SPOT FACTS

- The worldwide demand for electricity from power stations is doubling roughly every 20 years.

- World energy consumption grew by 1.8 percent in 1995, the highest annual growth rate since 1989.

- Waste water from the power station at Svartsengi, Iceland, is used to heat a lagoon to 70°C (158°F), and is used by bathers suffering from skin disorders.

- In 1882, Thomas Edison opened the first commercial power station, in Pearl Street, New York. His main reason for doing so was to sell more of his new electric lamps.

- The largest and most powerful steam turbines can each produce enough electricity to supply nearly 800,000 houses.

- The condensers for the steam turbines of a 2,000-megawatt power plant require more than 200 million liters (over 211 million quarts) of cooling water every hour.

- The high-pressure steam traveling from a power station boiler to the turbines has a temperature of about 500°C (932°F).

- Fuel-burning power plants have an average efficiency of about 35 percent: nearly two-thirds of the heat energy they produce is wasted.

- Combined heat and power (CHP) stations can use the "waste" heat from electricity generation to keep buildings warm.

- Some power-station boilers can be run on chicken droppings, old automobile tires, or wet sewage.

- The world's largest power plant is the hydroelectric station on the border between Brazil and Paraguay. It supplies 40 percent of Brazil's energy needs, and most of the energy needs of Paraguay.

- A large power station could power 5 trillion (5 million million) calculators.

- The main difference between the generator of a power station and a bicycle dynamo, apart from size, is that inside the dynamo there is a rotating permanent magnet, while inside the generator there is a rotating electric coil.

- A turbine works in a similar way to a windmill – the force of gas or steam (instead of wind) acts on vanes or blades.

- An automobile has a dynamo, or alternator, that creates a voltage of 12 volts to keep the battery charged. A power-station generator typically provides 22,000 volts.

- The United States has more nuclear reactors than any other country in the world. Its 109 reactors produce nearly 30 percent of the world's total power.

The generation game

The useful properties of electricity were first demonstrated by the British physicist Michael Faraday. In 1821, Faraday built the first electric motor. Ten years later, he developed the principle behind the electrical generator, which is basically an electric motor in reverse. By the 1860s, small electrical generators powered by steam engines were being used in many parts of Europe and the United States.

The first generating stations were built to supply electricity to large, individual buildings, such as hospitals and indoor markets. During the 1880s, demand for this new source of energy grew rapidly, and the first central generating stations were built. These were intended to supply electricity to the general public across a whole city district. In 1882, the American inventor Thomas A. Edison opened the famous Pearl Street power station in New York.

In order to produce electricity in commercial quantities, generators need to rotate at speeds of at least 1,000 rpm (revolutions per minute). Steam engines could just barely achieve that speed, and it meant operating to the limits of their capacity.

During the 1880s and 1890s, the British engineer Charles Parsons perfected the inward-flow steam turbine, which could produce speeds of up to 18,000 rpm. Parsons's invention was rapidly adopted by the new electrical industry, and the first large steam turbines were installed in a German power station in 1901. Since then, steam turbines have become our most important source of electrical power.

Most power stations produce electricity in a two-stage process. Fuel of whatever kind – whether it be coal, oil, gas, or uranium – is first used to produce steam. The steam is then used to drive a turbine and generator, which produce useful quantities of electricity.

▼ The generator room at the Paris Opera House, installed by Thomas Edison in 1887. Steam engines worked flat out to provide high speeds of rotation. The rotary power was transferred to the generators by a series of belts. Today's power plants have turbines and generators mounted in line, and sharing a common rotating shaft.

▼ The first electrical generator built by Michael Faraday. Electrical current was produced by a copper disk spinning between the poles of an electromagnet. Modern generators use coils of copper wire instead. The first generators produced direct current. The switch to alternating current was made during the 1890s.

Steam turbines

A steam turbine consists of a central shaft, or rotor, mounted horizontally within a cylinder. The outer surface of the rotor is equipped with a large number of angled blades that radiate like the spokes of a wheel. High-pressure steam is passed into the cylinder through a series of nozzles mounted around the inner surface of the cylinder.

As the steam enters the cylinder, it expands. The heat energy of the steam is the kinetic energy (the energy of motion) of the vapor molecules. This energy is transmitted to the shaft through the angled blades, which move when the steam hits them, causing the shaft to rotate. The speed of rotation depends on the temperature and pressure of the steam.

Steam turbines constructed from newly developed materials can operate at extremely high temperatures and pressures. The design of the angled blades is an important factor.

The simplest form of steam turbine is the impulse turbine, in which the blades are shaped like tiny cups. The steam releases its energy when it hits the turbine blades. The shaft rotates, but the steam loses most of its speed.

A reaction turbine, such as the one originally invented by Charles Parsons in the late nineteenth century, allows the steam to flow through a series of fixed and moving blades. As it passes through the blades, the steam continues to expand. This means that the steam energy is used more efficiently, and the flow of steam is twice as fast as in an impulse turbine. Modern steam turbines usually incorporate both types of blades in their design.

Still greater efficiency can be obtained by allowing the energy transfer to take place in a number of separate stages rather than all at once. This process, which is known as staging, is now used in all steam turbines.

▲ In a modern steam turbine, superheated high-pressure steam expands through progressively larger turbine blades. It is piped first into the high-pressure turbine, and then, after reheating, into the medium-pressure and then the low-pressure turbines. The steam condenses back into water in the condenser, from where it returns to the boiler.

▶ Steam turbines being assembled in a factory. In order to withstand constant exposure to high-temperature steam, the sets of blades have to be made from extremely tough, resistant materials.

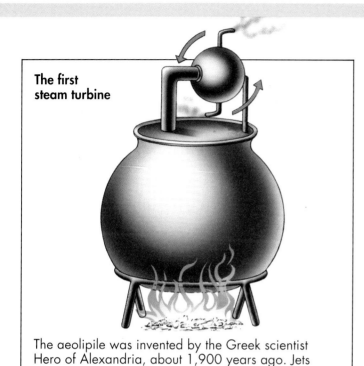

The first steam turbine

The aeolipile was invented by the Greek scientist Hero of Alexandria, about 1,900 years ago. Jets of steam caused the ball to rotate.

In the most common method of staging, the steam passes through a series of angled blades of progressively larger diameter. The steam drives the smallest set of blades first.

Large turbines take staging a step further. Two or three separate sets of turbine blades are mounted on the same shaft. Each stage is contained within a separate cylinder. The first stage usually has impulse blades and uses very-high-pressure steam.

After the steam has passed through the first stage, it is collected and reheated until it has sufficient energy to drive medium-pressure turbine blades. The steam is then piped into the low-pressure turbines. Turbines like these are known as compound turbines.

▼ A large compound steam turbine of the type found in most power stations. The many sets of steam-driven blades are all attached to one central shaft, which then turns the rotor inside the generator.

Power-station turbines: High-pressure turbine, Medium-pressure turbine, Low-pressure turbines, Turbine stages, Stator, Rotor, Generator, Exciter, Condenser, Air pumps, Reheated-steam valve, High-pressure-steam valve

Nuclear power

The CANDU reactor

A nuclear power plant has the same basic layout as a plant fueled by coal, oil, or gas. Water is heated in large boilers to produce steam. The steam is piped into turbines that turn generators, and is then cooled and condensed.

Although nuclear power plants do not produce smoke or fumes, there is always a danger that radioactive gases may leak into the environment. Within the power plant, the operators have to take strict precautions to prevent the accidental escape of radioactive substances.

The nuclear reactor itself is housed in a separate building with very thick concrete walls to absorb neutrons and other types of radiation. The reactor's coolant system transfers heat to heat exchangers, which produce steam. The heat exchangers are also located in the reactor building. They are fed by an outside supply of water, and the steam leaves through underground pipes. This cuts to a minimum the time that the water and steam are exposed to neutrons spreading out from the core.

Safety systems

If the chain reaction in the core was to go out of control, it would cause the reactor to overheat. A number of safety systems are in place to prevent this. When the temperature of the core starts to rise, control rods are automatically lowered into it to slow down the chain reaction. If necessary, they can be used to shut down the core completely. A secondary safety system, located above the reactor, can be used to drench the core with cold water. Apart from two major accidents – one at Three Mile Island in the United States and one at Chernobyl in the Ukraine – the nuclear power industry has had a very good safety record.

◀ The CANDU (Canada Deuterium/Uranium) reactor was first used in 1971 at the Douglas Point power station in Ontario, Canada. It is widely considered to be one of the safest and most efficient, and has been exported to several countries including Argentina and India. The reactor building consists of a cylinder of prestressed concrete, lined with additional shielding and reinforcement. The emergency drenching system is located just below the domed roof. Cold water for the steam condensers in the turbine room may be taken from the nearby river or lake. Some nuclear power plants take cooling water from the sea.

Key
1 Reactor building
2 Nuclear reactor
3 Heat exchangers
4 Water storage for emergency drenching
5 Drenching sprays
6 Control room
7 Steam in underground pipes
8 Storage for radioactive fuel
9 Turbines
10 Generator
11 Water intake for condensers
12 Pump house for cooling water
13 Water outlet pipes
14 Standby diesel generator

Wind and water turbines

Wind and water turbines work on the same principles as steam turbines, but operate at much lower pressures. The amount of energy available from running water depends on the vertical distance that the water falls. This distance is known as the head. The greater the head, the greater the energy provided by a given quantity of water. In places where the head is above 30 m (100 ft.), there is enough energy to create a high-pressure water jet. This can be used to generate electricity in a simple impulse turbine, such as the Pelton wheel.

Most hydroelectric power stations use a type of turbine developed by the American engineer James Francis in the mid-1800s. The Francis turbine is a reaction turbine. The angled blades are completely immersed in water, and are driven by the flow of the water through them, rather than by the impact of water against them. This enables the Francis turbine to operate with only a few meters of head, that is, using water with less pressure.

A Francis turbine consists of a single rotor, also called a runner, fitted with angled blades. The rotor is mounted within a spiral volute chamber. The turbine can be installed vertically or horizontally. In most cases, the rotor is mounted horizontally, and the electric generators are powered by a vertical shaft.

Where the head is too low for a reaction turbine, special low-head turbines can be used. One such design, the Kaplan turbine, uses a runner shaped like a ship's propeller. The Kaplan turbine can be installed within a volute chamber, or can be placed directly into the flow of water.

Wind turbines were developed, based on the traditional windmill, and many different designs are currently in use. Most installations use horizontal-axis turbines with rotors shaped like an airplane propeller. Vertical-axis wind turbines are less common, but they have the advantage that they can operate whatever the wind direction.

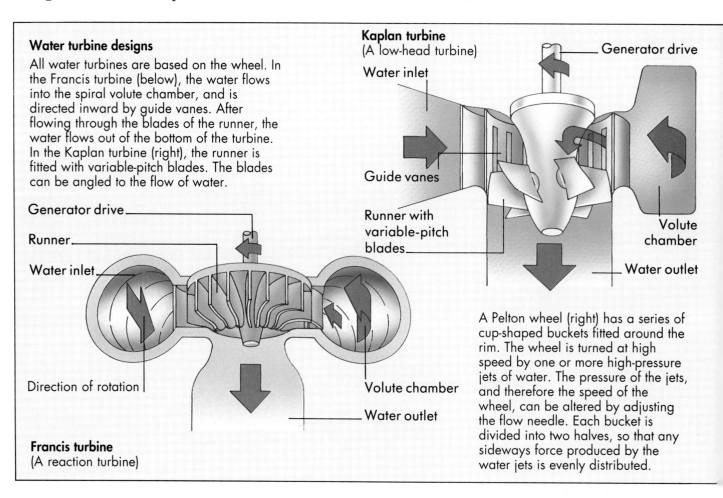

Water turbine designs
All water turbines are based on the wheel. In the Francis turbine (below), the water flows into the spiral volute chamber, and is directed inward by guide vanes. After flowing through the blades of the runner, the water flows out of the bottom of the turbine. In the Kaplan turbine (right), the runner is fitted with variable-pitch blades. The blades can be angled to the flow of water.

Francis turbine
(A reaction turbine)

Kaplan turbine
(A low-head turbine)

A Pelton wheel (right) has a series of cup-shaped buckets fitted around the rim. The wheel is turned at high speed by one or more high-pressure jets of water. The pressure of the jets, and therefore the speed of the wheel, can be altered by adjusting the flow needle. Each bucket is divided into two halves, so that any sideways force produced by the water jets is evenly distributed.

▼ The generator room at a hydroelectric power plant in the Snowy Mountains in Australia. The generators themselves are on the next level down, and the water turbines are on the next level below that. The power plant forms part of a mammoth hydroelectric and irrigation project. The project involved diverting the Snowy River back on itself so that it can be used to irrigate arid regions.

Pelton wheel (A simple impulse turbine)
Water jet
Generator drive
Buckets

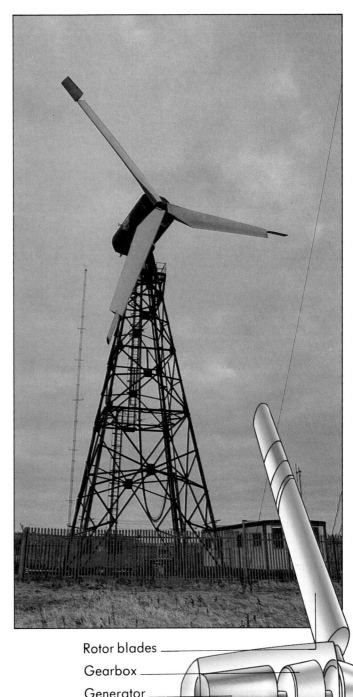

Rotor blades
Gearbox
Generator
Current flow

▲ A horizontal-axis wind turbine. Most designs use a three-bladed rotor, although some have only two. The main disadvantage of horizontal-axis turbines is that they can operate at maximum efficiency only when the wind is blowing straight at them.

Powerhouses

A power plant produces electric current in massive generators. These are often called turbogenerators, because they are driven by steam or water turbines. A more precise name is turboalternators, because they produce alternating current.

In power plants that use fossil or nuclear fuels, steam for the turbines is sent along insulated pipes from the reactor or boiler room. Turboalternators operate at a constant speed, which is adjusted by varying the pressure of the steam entering the turbines. In the United States, turboalternators typically operate at 3,600 revolutions per minute (rpm) to produce alternating current at 60 hertz (cycles per second). In Europe, they rotate at 3,000 rpm to produce current at 50 cycles.

The very high speeds at which they operate mean that turboalternators also produce large amounts of unwanted heat due to friction. Modern designs are filled with hydrogen gas under pressure, which enables the plant to lose heat more efficiently. Water is circulated through the outer casing in order to carry the unwanted heat away.

Electricity drawn from each turboalternator flows through three cables in order to produce a three-phase supply. The cables run under the powerhouse floor to the bus room. In the bus room, electricity from the various turboalternators is brought together at a central point. The bus system usually consists of a network of heavy copper bars or cables, separated by large insulators.

Individual sections of the bus system are separated by circuit breakers. If there is a fault in any part of the system, these automatically stop the flow of current and isolate the faulty section. The flow of electricity can also be regulated by heavy-duty switchgear, operated from the power plant's control room. Before electricity leaves the power plant, it passes through a series of transformers, which produce the very high voltages required for long-distance transmission.

In addition to the main turboalternators, most power plants also have smaller generators powered by gas turbines. These are used to produce additional electricity during periods when demand is especially high.

▲ The bus system and switchgear are protected against power surges by automatic circuit breakers.

◄ Supervisors in the control room constantly monitor the output of each turboalternator. They route the electricity through the bus system to the main transformers.

▲ In the final stage of electricity production, the transformers step up the current to very high voltages.

▼ The powerhouse of a coal-fired power plant. Steam for the lines of turboalternators is piped up from the boiler room below. Cables carrying electric current to the bus room are set into the floor. Overhead cranes are used to lift up the massive outer casings of the turboalternators when they need servicing and maintenance.

Generators

Rotating a coil of wire inside a magnetic field causes an electric current to flow through the coil. If the ends of the coil are connected to a split commutator, it produces direct current. Direct current is one that flows in one direction only. Alternating current is produced by connecting the coil to two separate slip rings. Every half turn of the coil reverses the direction of the current through the bulb.

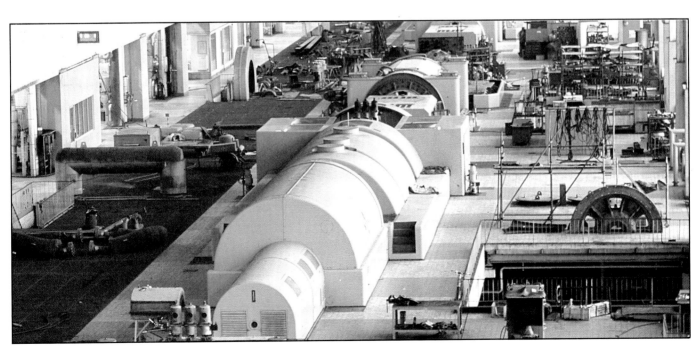

57

Using electricity

Electricity is instantly available at the touch of a switch. During the last 100 years or so, electricity has completely transformed human society. Electricity provides basic heating and lighting, and also powers a wide range of machines. In the developed countries, many homes now contain over a dozen different electrical appliances.

Power plants produce high-voltage electricity, which may be carried by power lines for hundreds of miles, for use in factories and houses. Household appliances, however, use much lower voltages, and during the distribution of electricity, the voltage is progressively reduced. Inside the home, electricity is carried by a number of different circuits.

SPOT FACTS

- By 1900, more than 2.5 million light bulbs were being used to light London.

- In the United States, some power lines carry electricity at 765,000 volts.

- The biggest-ever power failure happened in 1965, when an area of over 200,000 sq km (80,000 sq. mi.) in 8 northeastern U.S. states and Ontario was blacked out for up to 13 hours. Some 30 million people were affected.

- "Blackout" is a word used to describe power failure, whether accidental or planned. The term "brownout" is used to refer to a reduction of power.

- European electricity supplies in houses have a voltage of 230 volts. This is an average value, because the voltage varies between 0 volts and about 330 volts, 50 times every second.

- The first electric lamps used the light of a spark created between two carbon rods by a large battery. The lights were very bright, but they were noisy and the carbon rods had to be changed frequently.

- In 1879, a white hot filament, like the kind used in a modern light bulb, was first used in an electric lamp.

- The filament in a modern electric light bulb is made from tungsten, a metal with a very high melting point. Its working temperature is about 2,500°C (4,532°F).

- Tungsten metal in a filament lamp slowly evaporates into the gases in the bulb. The tungsten gradually gets thinner and thinner, until it reaches breaking point.

- Fluorescent tubes are about five times more efficient than filament lamps.

- An electric motor uses current through a wire to create a magnetic field. This interacts with another magnetic field to produce motion.

- An electric generator uses the motion of a wire through a magnetic field to create a current.

- In the United Kingdom, people use an average of six household batteries every year.

- No electric current flows through a bird standing on a wire. But a bird with a large wingspan can be electrocuted if its wings touch a live cable and an earthed pylon at the same time.

- A heart pacemaker gives small, regular electric shocks to the heart muscle to keep it beating steadily.

- In most countries, electricity consumption during the night is charged at a lower rate than during the day, because it is more efficient to keep a power station running than to heat it up every morning.

- The disadvantage of electric automobiles is that their batteries need to be recharged every 160–240 km (100–150 mi.). However, they are less polluting to the atmosphere than gasoline automobiles.

Transforming society

The single most important electrical device is the light bulb. Before its invention, the only forms of artificial light were oil and gas lamps. The first practical electric light bulbs were developed during the 1870s by Joseph Swan in Great Britain, and by Thomas Edison in the United States. Electric lighting was fairly cheap to install, and it was much more convenient, and safer than oil or gas. By 1900, electricity was in use in many parts of the world.

Other useful electrical appliances were soon developed. By 1890, electrical engineers had perfected the electric space heater, with long-lasting heating elements made from nickel-chromium alloy. Electric stoves and vacuum cleaners first went on sale in the 1890s, but were not widely used before the 1920s.

Since 1920, electricity has completely transformed daily life, and a wide variety of household appliances are now available. Electric irons, food processors, washing machines, air-conditioning units, refrigerators, and microwave ovens are now widely used.

Electricity also provided the energy for new forms of mass communication and popular entertainment. First radio and then cinema and television were made possible by the availability of electrical energy.

Industry too was transformed; electricity is now used in many basic processes, especially in the metalworking and chemical industries. More recently, electricity has allowed the rapid development of electronic computers and the revolution in information technology.

◀ The Crystal Palace was built in London during the age of gas and oil lamps. In 1885, it was lit by electricity and provided a dramatic advertisement for the new energy source.

▼ The electric light bulb, invented by Thomas Edison in 1879. The first light bulbs had a carbon filament made from cotton thread, which only lasted for about 40 hours. By 1900, there were longer-burning filaments made from metal wire. Modern bulbs use tungsten filaments.

Distribution

High-voltage power lines carry electricity over long distances. Energy is lost, mainly in the form of heat, as electricity travels along wires and cables. The use of very high voltages, typically ranging from 230,000 to 765,000 volts, reduces this energy loss to a minimum. The most widely used distribution method is by overhead power lines that are supported by metal towers, called pylons, or poles. The atmosphere provides cooling and most of the insulation. At points where the power line passes under the arm of a tower, it is held by a glass or ceramic insulator.

Underground high-voltage lines are much more expensive to install, and are normally used only over short distances. The power line may be wrapped in paper impregnated with insulating oil and is contained in an oil-filled metal pipe. Gas-filled pipes are also used.

When electricity is taken from power lines, the voltage is stepped down by banks of transformers at a substation. Most substations also contain circuit-breakers and switchgears.

Stepping down the voltage to the levels used by the customer usually requires a series of substations. Factories often use electricity at 10,000 volts or more, which is much too high for domestic use. In cities, a series of local substations may provide electricity for distribution to houses, apartment blocks, and stores. Large office buildings usually take electricity at a higher voltage than they can use directly, and so have a separate substation on the premises to step down the voltage.

In isolated regions, a power plant may be constructed to provide electricity directly to a town or large factory. In most cases, however, all the power plants in a country are linked into a national grid of power lines. Output from power plants can then be matched more closely to demand in different places, and electricity can even be traded between different national grids. This normally takes place by overhead power line. Great Britain, however, imports some of its electricity from France along an undersea cable.

▼ Electricity is generated as alternating current at a power plant, where transformers step up its voltage to several hundred thousand volts. Power lines carried by tall towers transmit the high-voltage current to local substations, where transformers step down its voltage. Factories may take the electricity at over 10,000 volts. Homes use electricity usually at 115 or 230 volts.

From power station to consumer

- Coal train
- Furnace and boiler
- Cooling tower
- Turbo-alternator
- Cooling water
- Transformers and switchgear
- High-voltage power lines
- Towers

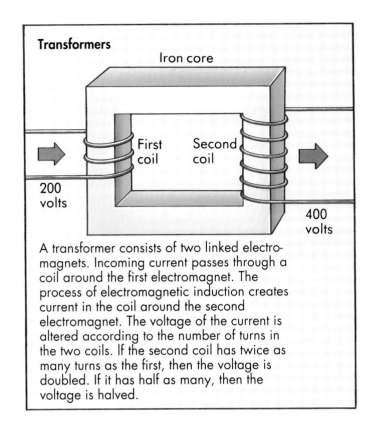

Transformers

A transformer consists of two linked electromagnets. Incoming current passes through a coil around the first electromagnet. The process of electromagnetic induction creates current in the coil around the second electromagnet. The voltage of the current is altered according to the number of turns in the two coils. If the second coil has twice as many turns as the first, then the voltage is doubled. If it has half as many, then the voltage is halved.

▲ A tower carrying high-voltage power lines, which here consist of sets of three cables. They have an aluminum conductor and are reinforced with steel.

61

Heating and lighting

Electricity gives off heat when it meets resistance to its flow. Useful quantities of heat can be produced by passing electricity through a coil of thin wire. A thin wire has a higher resistance to electricity than a thick wire. Winding the wire into a coil concentrates the heating effect into a small area. Coils used for heating are called heating elements and are found in many household appliances.

An electric light bulb works by heating a coil of very thin wire, known as a filament, until it glows white hot. The first light bulbs contained a vacuum to prevent the filament from burning up. Modern bulbs are filled with an inert (chemically inactive) gas such as argon. A light bulb produces light across the whole range of the visible spectrum, but is very inefficient in its use of electricity. Only about 6 percent of the energy is released as useful light; the rest is lost as heat.

Discharge lamps are much more efficient at turning electricity into light. A discharge lamp consists of a glass tube containing a gas or vapor that conducts electricity. Metal contacts at each end of the tube allow current to pass through the gas. The flow of electricity causes the gas atoms to agitate in the tube, and as a result they give off light.

Different gases produce different colors of light. Sodium vapor produces orange light, and mercury vapor produces blue light. Ordinary discharge lamps are widely used for street-lighting, but the color of the light makes them unsuitable for use in the home or office.

Discharge lamps filled with neon gas can produce a wide variety of colors and are used in illuminated signs, often for advertising purposes. Other types of electric lamp are designed to produce infrared light, and are often used as bathroom heaters.

Fluorescent light tubes

Fluorescent tubes are special discharge lamps containing mercury vapor. They produce white light by means of a coating of phosphors on the inside of the tube. Besides visible blue light, mercury vapor gives off an invisible ultraviolet light. The phosphors absorb this and give off a white light that is suitable for most purposes.

Fluorescent tubes use much less electricity than ordinary light bulbs and last much longer. Small fluorescent tubes are now being incorporated in "economy" light bulbs that can be substituted for the ordinary type.

▶ The city of Boston by night. Electric lighting has greatly extended the working day, and offices and other buildings can now operate around the clock. Outside, powerful electric floodlighting also enables open-air activities, such as football games and other sporting events, to take place in the evening.

The fluorescent tube

◀ Fluorescent tubes produce useful light in a two-stage process. Electrical contacts at each end of the tube allow current to pass through the mercury vapor, producing ultraviolet light. The phosphors on the inner surface of the tube absorb the ultraviolet light energy, and emit light across the whole range of the visible spectrum. Fluorescent tubes are more economical than filament lamps.

Transformers

A transformer consists of two linked electromagnets. Incoming current passes through a coil around the first electromagnet. The process of electromagnetic induction creates current in the coil around the second electromagnet. The voltage of the current is altered according to the number of turns in the two coils. If the second coil has twice as many turns as the first, then the voltage is doubled. If it has half as many, then the voltage is halved.

▲ A tower carrying high-voltage power lines, which here consist of sets of three cables. They have an aluminum conductor and are reinforced with steel.

Heating and lighting

Electricity gives off heat when it meets resistance to its flow. Useful quantities of heat can be produced by passing electricity through a coil of thin wire. A thin wire has a higher resistance to electricity than a thick wire. Winding the wire into a coil concentrates the heating effect into a small area. Coils used for heating are called heating elements and are found in many household appliances.

An electric light bulb works by heating a coil of very thin wire, known as a filament, until it glows white hot. The first light bulbs contained a vacuum to prevent the filament from burning up. Modern bulbs are filled with an inert (chemically inactive) gas such as argon. A light bulb produces light across the whole range of the visible spectrum, but is very inefficient in its use of electricity. Only about 6 percent of the energy is released as useful light; the rest is lost as heat.

Discharge lamps are much more efficient at turning electricity into light. A discharge lamp consists of a glass tube containing a gas or vapor that conducts electricity. Metal contacts at each end of the tube allow current to pass through the gas. The flow of electricity causes the gas atoms to agitate in the tube, and as a result they give off light.

Different gases produce different colors of light. Sodium vapor produces orange light, and mercury vapor produces blue light. Ordinary discharge lamps are widely used for streetlighting, but the color of the light makes them unsuitable for use in the home or office.

Discharge lamps filled with neon gas can produce a wide variety of colors and are used in illuminated signs, often for advertising purposes. Other types of electric lamp are designed to produce infrared light, and are often used as bathroom heaters.

Fluorescent light tubes

Fluorescent tubes are special discharge lamps containing mercury vapor. They produce white light by means of a coating of phosphors on the inside of the tube. Besides visible blue light, mercury vapor gives off an invisible ultraviolet light. The phosphors absorb this and give off a white light that is suitable for most purposes.

Fluorescent tubes use much less electricity than ordinary light bulbs and last much longer. Small fluorescent tubes are now being incorporated in "economy" light bulbs that can be substituted for the ordinary type.

▶ The city of Boston by night. Electric lighting has greatly extended the working day, and offices and other buildings can now operate around the clock. Outside, powerful electric floodlighting also enables open-air activities, such as football games and other sporting events, to take place in the evening.

The fluorescent tube

◀ Fluorescent tubes produce useful light in a two-stage process. Electrical contacts at each end of the tube allow current to pass through the mercury vapor, producing ultraviolet light. The phosphors on the inner surface of the tube absorb the ultraviolet light energy, and emit light across the whole range of the visible spectrum. Fluorescent tubes are more economical than filament lamps.

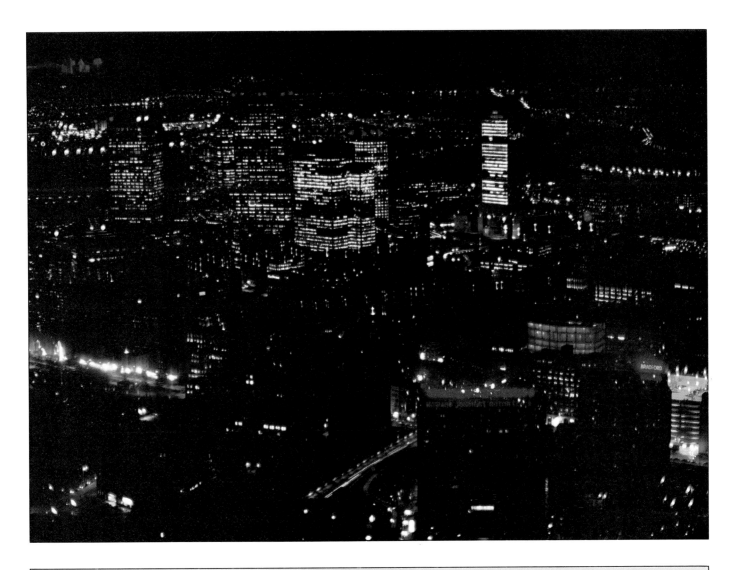

Heating by radio

A microwave oven heats food by means of high-frequency radio waves, known as microwaves, which have a very short wavelength. The microwaves are produced by a device known as a cavity magnetron. As the microwaves pass through an item of food, they cause its molecules to vibrate, and this vibration produces heat. The cavity magnetron was developed during World War 2 in order to produce tightly focused beams of microwaves for the first radar systems. Although the microwaves are said to form a continuous beam, the magnetron is in fact switching on and off about 10,000 times every second. The great advantage of microwave ovens is that they heat food extremely quickly.

Circuits and wiring

The electrical wiring inside a house consists of a number of ring (or loop) and branch circuits. The supply of electricity coming into the house from the local substation goes to a joulemeter and is then split into the different circuits at the service panel, which has fuses to protect against the fire hazard caused by excessive current. The circuits use the voltage supplied by the service wires coming into the house, but are designed to carry different loads of electric current. Each circuit has a separate fuse that melts if the circuit becomes overloaded. Some modern houses now use circuit breakers in the service panel instead of fuses in the event of an overload to the system.

A house has several circuits. There will be at least one circuit for the outlets, and at least one for the lights. Generally, there are also separate circuits for heavy-duty appliances such as electric stoves or water heaters. A circuit built into the ceiling supplies electricity to individual light fixtures. Light switches are usually wall mounted.

Dangers of electricity
Electricity is dangerous. The voltage of electricity supplied to a house is high enough to produce a current that can kill people. Do not attempt to investigate or experiment with any of the electrical fittings in your home.

Domestic wiring

▼ Electricity used in the home comes from the local substation. The service panel, with fuses or circuit breakers, is a safety device used to isolate the different circuits from each other; it does not alter the voltage of the electricity in any way.

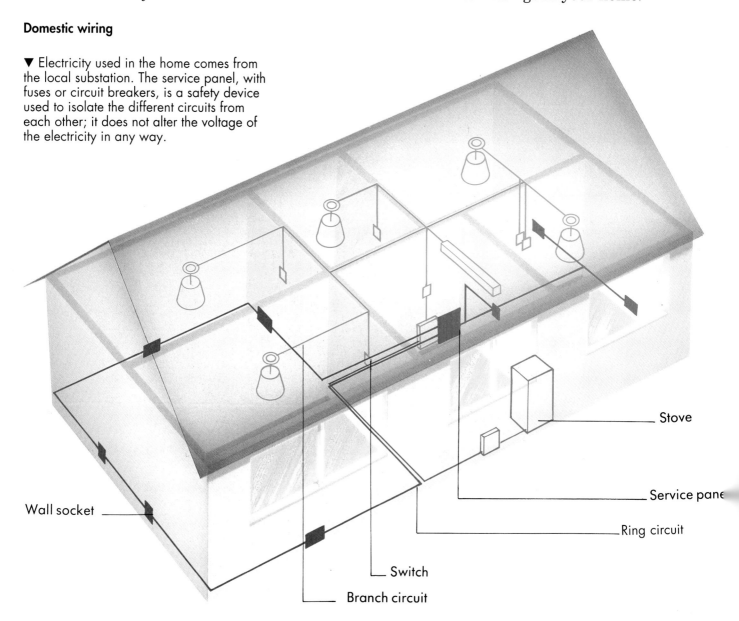

64

Motors

Apart from heating elements and lights, the most useful electrical devices are electric motors. Motors of many different sizes are built into many household appliances. Some use the motor's power directly, some indirectly.

Electric drills and food blenders make direct use of the motor's rotating shaft. Vacuum cleaners and hairdryers use motors to turn a fan, which creates a flow of air. This can be used to pull air into the appliance, or blow air out.

Water pumps driven by electric motors are found in washing machines, shower units, and central heating systems. Refrigerators also use electric motors; in these appliances they circulate special coolant liquids.

Electric motor

A direct-current motor consists of a coil of wire attached to a shaft and surrounded by a simple magnet. When electricity flows through the coil, the effect of the magnetic field causes the coil to rotate. The supply of electric current to the coil is controlled by a split-ring commutator and brushes.

▼ A hairdryer uses electricity in two different ways. A small electric motor drives a fan, and heat is produced by heating elements in the form of wire coils. As a safety precaution, the dryer is provided with a thermostat (heat-sensitive bimetallic switch). If the dryer overheats, the switch opens and breaks the electrical circuit.

Engines

Engines are devices that burn fuel and provide mechanical power. Steam engines and gasoline engines both produce power through the up-and-down, or backward and forward, motion of a piston inside a cylinder. Engines that operate in this manner are called reciprocating engines.

Steam engines burn fuel separately to produce steam in a boiler. They are therefore known as external-combustion engines. Gasoline engines burn fuel inside the engine itself, and are internal-combustion engines.

SPOT FACTS

- At normal atmospheric pressure, steam has a volume at least 1,600 times greater than that of the same mass of water.

- One of the most powerful gasoline engines ever built had 28 cylinders arranged in four circles of seven, and was used in American bomber aircraft during World War 2.

- In 1689, Denis Papin (1647–c.1712) tried to make a steam engine. It did not work, but it was good for cooking food. He had invented the pressure cooker by accident.

- Papin also tried to build an engine powered by gunpowder. That did not work either, but it was the first attempt at an engine that worked by combustion inside a cylinder – the process known as internal combustion.

- The first working internal-combustion engines, developed in the nineteenth century, ran on gas.

- In the nineteenth century, small steam engines could be bought to power household machines: for example, a steam-powered sewing machine.

- The steam engines built by James Watt had an efficiency of about 1 percent. They could convert about 1/100 of the total energy available from the burning fuel into useful work. A modern automobile engine has an efficiency of about 30 percent.

- Automobiles of the future may be powered by fuel cells that use hydrogen and oxygen to generate electricity to turn a motor. They are twice as efficient as engines that burn gasoline or diesel oil.

- A typical automobile traveling some 30,000 km (over 18,000 mi.) uses the same amount of oxygen that a typical tree produces in 100 years.

- There are about 600 million road vehicles in the world.

- A "high performance" automobile accelerates at about 6.5 km (4 mi.) per second. A sprinter's acceleration at the start of a race peaks at 16 km (10 mi.) per second.

- Every night, when you are asleep, some 30,000 new automobiles are made worldwide.

- The average automobile emits four times its own weight of carbon dioxide every year.

- The axle of a fast-moving automobile turns at about 1,000 rpm. The engine crankshaft turns at anything between 0 and about 10,000 rpm. The crankshaft must be connected to the axle using gears.

- Diesel engines of over 3,000 horsepower are used to power many modern locomotives.

- In a gas engine, an electric spark ignites the fuel. In a diesel engine, the air in the piston is heated by compressing it. The temperature is high enough to ignite the fuel.

- Because the air in the pistons of a diesel engine is compressed so much, the engine must be made of metal that is thick enough to be sufficiently strong. This is why diesel engines are much heavier than gas engines.

- Diesel engines are often used in trucks and buses in the United States, and in automobiles in Europe.

The age of steam

If steam condenses within a closed container, a partial vacuum is created. The first practical steam-powered device, invented by Thomas Savery in 1698, was a vacuum pump.

In 1712, another Englishman, Thomas Newcomen, built the first true steam engine, which used a piston within a vertical cylinder. Steam entering at the bottom of the cylinder forced the piston up. The steam was then condensed by a spray of cold water. The resulting vacuum allowed the piston to fall, ready to be driven up again by more steam. A connecting rod transferred the up-and-down motion to a hinged beam, hence the name "beam engine."

Newcomen's engines produced a steady pumping action, and were widely used in coal mines. But they were slow and inefficient.

Between 1769 and 1790, the Scottish inventor James Watt made a number of improvements, making the steam engine flexible and efficient. Watt's first improvement was the separate steam condenser, which allowed the cylinder to remain hot. He also invented double action, in which both the up-stroke and the down-stroke were steam-powered. Another improvement of his was to connect the piston to an offset flywheel, which converted the up-and-down motion into rotation.

▲ A contemporary print showing an industrial steam engine of the mid-1800s. The right-hand arm of the beam (d) was pulled down when steam expanded against a piston in the steam cylinder (a) and was condensed in a condenser (g).

▶ Savery's steam pump used two boilers side by side and was very unsafe. Water was alternately boiled and condensed in each of the two boilers. The vacuum that was produced drew water up a pipe and through a one-way valve.

Steam engines

During the 1800s, the steam engine was further transformed. Vertical cylinders were replaced by horizontal ones, and better engineering techniques allowed high-pressure steam to be used with safety. This meant improved boilers, with steam being heated to progressively higher temperatures in a series of tubes within tubes. As a result, steam engines became smaller and produced more power.

Some designs recycled steam through a series of expansion chambers and condensers, thus making maximum use of steam's heat energy. Such engines tended to be both large and heavy, and were mainly used in factories.

Other designs abandoned bulky condensers and discharged steam straight into the atmosphere once it had been used. These smaller and lighter engines were used mainly in steam locomotives.

A steam locomotive is a self-contained unit that produces enough power to pull heavy loads along railroad lines at high speeds. Fuel, usually coal, is burned in the firebox to produce heat for the boiler. Steam locomotives normally have two cylinders, one on each side. The movement of the pistons in the cylinders is transferred to the driving wheels through a series of connecting rods and cranks.

Steam locomotive

obsolete with the development in the 1880s of the gasoline-engine automobile. Steam-powered traction engines, however, were used on farms and in road building for many years.

Stationary steam engines of all sizes were used throughout the 1800s to produce power in factories in Europe and North America. But by 1900, large steam engines were being replaced by more efficient steam turbines, and smaller ones by internal-combustion engines, such as gas engines.

▲ In a locomotive, hot gases from the firebox are drawn through the boiler tubes and boil the water. The steam is fed into the cylinders to drive the pistons, which are connected by rods to the drive wheels.

◀ A powerful steam traction engine, of the type used on farms from the late 1800s. They were used to pull plows and power machinery such as threshers.

The first steam locomotive, designed by the English engineer Richard Trevithick, appeared in 1804. By then, steam power was already being used for transportation on water. A paddleboat called the *Charlotte Dundas* began operating in Scotland in 1801. The first crossing of the Atlantic by a ship using steam power came in 1819. Later, steam carriages and automobiles enjoyed popularity, but they became generally

Double action

In a double-acting steam engine, steam is introduced to each side of the piston alternately. In the diagram (top), the slide valve allows steam in on the right of the piston, and the steam forces it along the cylinder. The valve then moves across (above), allowing steam in on the left, and the spent steam escapes. James Watt introduced double action in 1781.

Gasoline engine

The gasoline engine is an internal-combustion engine. Fuel is burned inside a cylinder to provide the energy to drive a piston. Only one in four of the piston strokes produces power, and for this reason gasoline engines are known as four-stroke engines.

The four-stroke engine was invented in 1876 by the German engineer N. A. Otto. The Otto engine used coal gas as fuel. In 1885, another German, Gottlieb Daimler, invented the four-stroke gasoline engine. Many improvements were made by other engineers, and by 1900, the gasoline engine had acquired all the main features that it has today.

Gasoline engines burn gasoline vapor in a ratio of about 14 parts air to one part gasoline. The mixture is compressed by a piston and is then ignited by an electric spark. The combustion, or explosion, pushes the piston down. The downward motion is transformed into rotation by the crankshaft.

Because of the four-stroke cycle, gasoline engines operate most efficiently when they have four or more cylinders operating in sequence. One of the cylinders is always delivering a power stroke, which means that the crankshaft is constantly receiving power. The rotating crankshaft transmits the useful power produced by the engine. Most of this power is used to perform work, for example propelling an automobile.

Gasoline engines are mainly used in automobiles and motorcycles. Many automobile engines have four cylinders, and larger engines often have multiples of four. The cylinders can be arranged in a line, or in a V-shape. Engines are often described as V-4, V-6, V-12, and so on. Motorcycles often have one or two cylinders.

Gasoline engines are capable of sustained operation at high speed. When producing maximum power, the pistons may travel up and down the cylinders 175 times every second.

◀ A high-performance racing car cornering at high speed. Many high-performance engines use superchargers or turbochargers to compress the air, so that a greater quantity of fuel can be burned at each stroke, thus producing more power.

Four strokes

The first stroke in the cycle (the intake stroke) draws the gasoline/air mixture into the cylinder. The second stroke (the compression stroke) compresses the mixture, which is ignited when the piston is at the very top of the up stroke. During the third stroke (the power stroke), the piston is driven down by the rapidly expanding gases produced by combustion. The fourth stroke (the exhaust stroke) forces exhaust fumes through the exhaust valve, and the cylinder is ready to repeat the cycle. Only the power stroke drives the piston. During the other three strokes, the pistons are driven by the rotation of the crankshaft.

1 Intake stroke: Fuel mixture enters.

2 Compression stroke: Mixture is compressed.

The gasoline engine

Key
1 Air intakes
2 Air chamber
3 Fuel line
4 Fuel injector
5 Piston
6 Exhaust valve
7 Intake valve
8 Valve rocker arm
9 Camshaft
10 Crankshaft
11 Chain drive for camshaft
12 Exhaust manifold
13 Oil pan
14 Flywheel
15 Distributor
16 High-voltage leads
17 Spark plug connector

▲ Cutaway drawing of a V-4 gasoline engine, typical of those used in medium-sized automobiles. The cylinders are arranged in two rows of three, set at about 120° to each other. Air enters the intakes and passes through the air chamber to the intake valves. Gasoline is pumped along the fuel line and is injected into the air flow and vaporizes. After combustion, exhaust fumes pass through the exhaust valves and into the exhaust manifold. Most of the power produced by the engine is transmitted by the crankshaft to the flywheel at the back of the engine. At the front, the crankshaft also drives both the cooling fan and the overhead camshaft. The camshaft operates the valves by means of a series of rods connected to spring-loaded rocker arms.

Engine systems

In addition to fuel and air, a gasoline engine requires electricity, oil, and water. These are supplied by different systems.

The fuel system pumps liquid gasoline from the fuel tank and distributes it in controlled quantities into the air flow to the cylinders. Many engines regulate the flow of gasoline with mechanical devices known as carburetors. Some engines use electronically controlled fuel injectors. The engine's breathing system draws in air through a filter, and channels the air flow to the intake valves.

The exhaust system takes hot gases from the cylinders through the exhaust manifold and out of the engine. In most cases, a muffler is located at the end of the exhaust system. In many developed countries, it is a requirement that exhaust gases pass through a catalytic converter to reduce the amount of pollution emitted.

The engine's lubrication system ensures that all moving parts are coated with a thin film of oil. This reduces friction. The main oil reservoir is the oil pan, which is located at the base of the engine beneath the crankshaft.

Fuel system

▶ The fuel injector forces the correct amount of vaporized gasoline into the air flow just above the intake valve.

Cooling system

▼ The spring-loaded valves are opened and closed in the correct sequence by an overhead camshaft. This is driven by a chain connected to the crankshaft.

Valve mechanism

◀ The cooling system circulates water through the engine casing and radiator. If required, some of the waste heat can be directed into the passenger compartment. The operating temperature of the engine is regulated by a thermostat.

▶ All the moving parts in an engine receive constant lubrication. A pump draws oil from the oil pan and forces it under pressure through channels to the engine bearings on the crankshaft and to the camshaft. Other areas get splashed with oil. Despite lubrication, some engine wear takes place. For this reason the lubrication system incorporates a filter to remove solid fragments before they do damage.

▼ A spark plug creates an electric spark across a tiny gap between two metal contacts, or electrodes. The inner electrode is connected with the plug lead and receives high-voltage current from the ignition coil via the distributor. The outer, or ground, electrode is connected to the plug casing. The two electrodes are insulated from each other by a thick ceramic layer.

Some engines are designed to be air-cooled. They pass unwanted heat directly into the atmosphere. Air-cooled engines, such as many motorcycle engines, usually have a series of cooling fins around the cylinder casing.

Most gasoline engines use both air and water to carry heat away from the cylinders. Water circulates through the engine and radiator inside a sealed system. Heat from the engine heats the water, which is then cooled inside the radiator. A stream of air is drawn through the radiator by a fan mounted at the front of the engine. The fan is usually turned by a belt attached to the crankshaft.

The electrical system is based around a low-voltage (12-volt) battery. The battery supplies current to the ignition coil, which produces the high voltage needed by the spark plugs. High-voltage current is supplied to each spark plug in turn by the distributor, which is a kind of high-speed rotary switch. The battery is kept charged by a generator driven by the crankshaft. The electrical system may also be used to drive the fuel, oil, and water pumps.

Other engines

Other types of piston engine share the same basic design as the gasoline engine, but differ slightly in the details of their operation. Some, such as the diesel engine, are designed to use a different type of oil-based fuel.

The diesel engine was invented by the German engineer Rudolf Diesel in 1893. Like the gasoline engine, it is a four-stroke engine. It differs because it has no separate ignition system to produce a spark. It is self-igniting, and uses the principle that air heats up when it is compressed. The action of the piston in a diesel engine compresses the air so that it becomes hot enough to ignite the fuel.

Diesel engines are more efficient than gasoline engines, but tend to be heavy. They are mainly used in large vehicles, trucks, buses, agricultural vehicles, and railroad locomotives. Small diesel engines are found in some cars. Diesel engines are also widely used to power small-scale electrical generators, such as emergency back-up generators in hospitals.

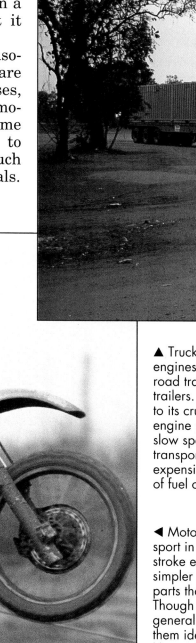

▲ Trucks powered by large diesel engines are used in Australia to pull road trains, consisting of several trailers. Once the road train gets up to its cruising speed, the diesel engine performs very efficiently. At slow speeds, however, this form of transportation is extremely expensive because of its high level of fuel consumption.

◀ Motorcycle riding is a popular sport in many countries. Two-stroke engines have a much simpler design and fewer moving parts than four-stroke engines. Though not as powerful, they are generally more reliable. This makes them ideal for young riders who like to do their own servicing and maintenance.

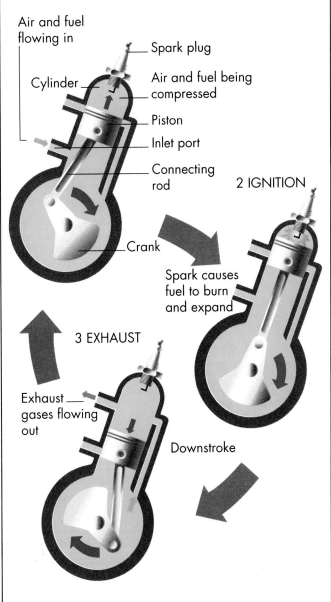

Two-stroke engines

Intake and exhaust are controlled by the piston. The fuel mixture of gasoline, oil, and air enters the inlet port and then transfers to the cylinder. The upstroke of the piston closes the transfer port; the fuel mixture is compressed and then ignited by a spark plug. During the downstroke (the power stroke), exhaust gases leave through the exhaust port, and fresh mixture flows in through the transfer port.

1 UPSTROKE

Air and fuel flowing in — Cylinder — Spark plug — Air and fuel being compressed — Piston — Inlet port — Connecting rod — Crank

2 IGNITION

Spark causes fuel to burn and expand

3 EXHAUST

Exhaust gases flowing out — Downstroke

The two-stroke engine is a gasoline engine that produces power every other stroke. Two-stroke engines have no separate lubrication system, and lubricating oil has to be mixed into the gasoline. Combustion is less efficient, and large amounts of exhaust fumes are produced.

Some gasoline-engine designs do not use a piston within a cylinder. Rotary engines, such as the Wankel engine, use a flat, three-cornered rotor that turns around a central shaft. It revolves in a chamber shaped like a figure eight. As it revolves, it creates spaces in which the four stages in the four-stroke engine cycle take place. These four stages are intake, compression, power, and exhaust.

The basic gasoline engine can also be adapted to run on other fuels. Some buses and trucks run on natural gas or methane. In Brazil there are automobiles that are powered by alcohol.

75

Jets and rockets

Jets and rockets produce power in the form of direct thrust. They burn fuel and concentrate a stream of exhaust gas through a nozzle at the back of the engine. This backward flow of hot gas causes the engine to move forward.

Jet engines breathe air, and are now used in all types of aircraft and helicopters. Rockets carry their own oxygen supply and can produce far more power than jet engines, but can burn fuel for only a fairly short period of time. During the period of combustion, rockets can achieve very high accelerations so that they can travel into space.

SPOT FACTS

- In 1947, the American pilot Charles "Chuck" Yeager became the first person to fly faster than the speed of sound in the rocket-powered X-1.

- Modern jet engines are at least 10 times more powerful than the biggest gasoline engines.

- The Russian *Energiya* rockets weigh 2,000 metric tons on the launchpad. Most of this is fuel. The satellites they carry weigh only 90 metric tons.

- The *Energiya* rocket has a lift-off thrust of nearly 3,500 metric tons.

- A modern jet engine can exert a force that would be big enough to lift a load of 50 metric tons vertically upward.

- The expanding gases traveling out of the back of a jet engine can have speeds of more than 2,000 km/h (1,242 mph).

- The direction of the thrust that drives an aircraft is opposite to the direction of the stream of high-speed gas created by its engines.

- A typical turboprop jet engine has the same power as approximately 2,000 horses (2,000 hp).

- The casing of a jet engine must be lined with "bullet proof" material in case a piece of blade breaks off and flies out from the engine.

- It costs the United States $100 million to launch a rocket carrying a big satellite.

- Helicopters and vertical take-off aircraft push hot air and gases directly downward so they can hover.

- The "wings" of a helicopter are called rotors. The rotors have a wing-shaped cross-section, and they spin at about 300 rpm.

- A Harrier "jump jet" uses swiveling nozzles to direct the exhaust gases from its jet engine downward as it takes off and lands vertically. It swivels the nozzles through 90° to produce horizontal thrust so that it can fly normally.

- Computers are used to control the amount of flow of a rocket's propellant. If the thrust is not correct, the rocket may go off-course and into wrong orbit.

- The space shuttle contains 49 rocket engines, 23 radar and radio antenna, and five computers.

- The space shuttle's reusable boosters carry over 700,000 kg (1.5 million lbs.) of liquefied propellant.

- The hydrogen fuel in a rocket is stored in liquid form at -253°C (-423°F). This is the temperature at which hydrogen becomes liquid.

- When hydrogen and oxygen burn together they produce water. The "exhaust" gas that drives a rocket forward is mostly water.

- A solid-fuel rocket is like a firework – there is no way to control the rate of burning once it is lit.

- The solid-fuel rockets on a European *Ariane* rocket last for just two minutes. After that, all of the thrust of the rocket must come from its liquid fuel.

The jet age

Until the 1950s, most aircraft were propeller-driven. The rotary motion needed to turn the propellers was provided by high-performance gasoline engines. The most powerful engines could propel a single-seater aircraft up to about 660 km/h (410 mph). Higher speeds required a different kind of engine, and by the 1930s many aircraft engineers were thinking about jet propulsion.

The invention of the jet engine is generally credited to Frank Whittle, an officer in the British Royal Air Force (RAF). Whittle patented the basic design of the jet engine in 1930, but independent research was also taking place in other countries. The first jet-powered flight was made by an experimental German aircraft, the Heinkel He-178, in August 1939.

Whittle's design was for what we today call a turbojet. Kerosene fuel is burned inside a stream of compressed air. The expanding exhaust gases drive a turbine before leaving the engine and providing thrust. Such jet engines are thus gas-turbine engines.

Toward the end of World War 2, Whittle's engine was used in the Gloster Meteor, the first jet aircraft to enter regular service. The Meteor, which became operational in 1944, was a twin-engined jet fighter, with one engine in each wing. The first jet-propelled passenger aircraft was the De Havilland Comet, which made its maiden flight in 1949, and entered service in 1952. Other passenger jets, including the Boeing 707, soon followed. By the 1960s, jets had largely replaced gasoline-engined aircraft.

The main advantage of the jet engine is that it produces far more power than a gasoline engine of the same size and weight. Jet engines can therefore propel larger aircraft at higher speeds over greater distances. Another advantage is they are mechanically simpler, producing rotary motion directly without the need for connecting rods and cranks. They also run more smoothly, with less vibration. And whereas the efficiency of propellers decreases with altitude, that of jets increases. Many aircraft now use turboprop and turbofan engines.

▲ A modern turbofan engine, of the type used to power most of today's airliners. It differs from the turbojet of the Whittle design in having a huge fan in front. The fan not only directs air through the engine, but also directs it around the engine. For this reason it is sometimes called a by-pass turbojet.

▼ Frank Whittle in 1944, the year that Britain's Gloster Meteor jet fighter went into service with the RAF. The first flight of a plane with a Whittle jet engine was in 1941, two years after the first jet plane, the He-178.

Jet engines

The simplest form of jet engine is the ramjet, which has no compressor or turbine. A ramjet relies on the speed of its own motion through the atmosphere to supply air quickly to the front of the engine. Fuel is burned at the center of the engine, and the expanding gases are directed through a nozzle at the rear.

The main disadvantage of the ramjet is that it cannot operate when the engine is not moving. Ramjets are therefore used only in missiles launched by a rocket motor. Once the missile is moving, the ramjet takes over.

A turbojet uses a rotary compressor at the front of the engine to provide high-pressure air to the combustion chamber. The exhaust gases pass through a turbine before leaving through the exhaust nozzle. The turbine provides just enough power to drive the compressor by means of a central shaft.

The power output of a turbojet can be boosted for short periods by afterburning. Additional fuel is burned in a second combustion chamber, which is located between the turbine and the exhaust nozzle.

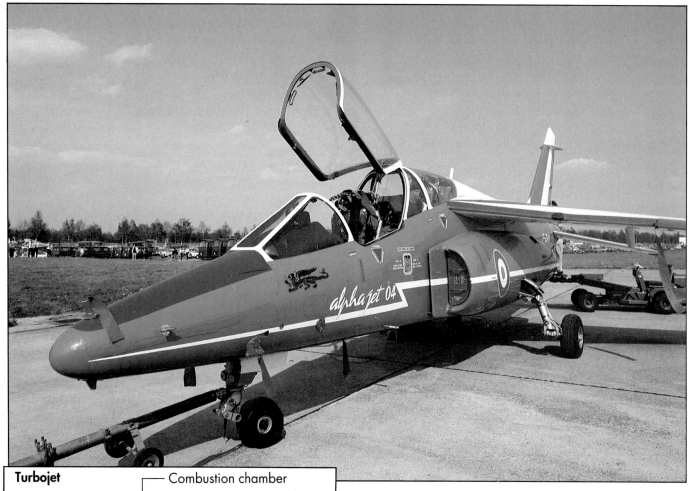

▲ Turbojets are now mainly used in high-performance aircraft such as this two-seater Alphajet. The air intakes are situated on either side of the fuselage just below the cockpit canopy.

◀ The turbojet consists of a central combustion chamber with a compressor in front and a turbine behind. Modern engines use a multistage compressor to force air into the combustion chamber.

▲ The turbofan engine has a propellerlike fan in front, which forces air around, as well as through, the compressor and combustion chamber. The fan is driven by a turbine in the jet exhaust. The bypass air joins the gases in the exhaust to produce thrust.

◀ With wing flaps and landing gear down, an Airbus A310 airliner of Switzerland's national airline, Swissair, comes in to land. The plane has two turbofan engines.

▶ Four turboprop engines power this curious-looking aircraft, known as the Super Guppy. They give it a maximum speed of about 460 km/h (nearly 290 mph).

▼ In the turboprop engine, the propeller is mounted outside. It is driven, through reduction gearing, by a separate set of turbines in the jet exhaust.

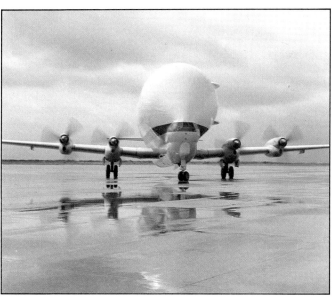

A turboprop engine operates in much the same way as a turbojet, but has two sets of turbines. One set drives the compressor, while the other drives a propeller mounted at the front of the engine. Although most propulsive thrust comes from the propeller, some comes from the gases exhausting from the turbines.

Turboprops are more efficient than turbojets at low speeds, and are widely used in multi-engined transport aircraft. The rotary motion that this type of engine produces also makes it suitable for helicopters. In the helicopter, it is known as a turboshaft engine.

The most common engine used in passenger-carrying airliners is the turbofan, which is a development of the turboprop. In a turbofan engine, the propeller, or fan, is mounted inside the engine casing in front of the compressor. Part of the air drawn in by the fan is directed into the compressor and combustion chamber. But most is directed around them, straight into the jet exhaust. About 75 percent of the engine thrust is produced by the fan. Turbofans are also called ducted-fan or bypass turbojet engines. They are quieter than turbojets and turboprops and consume less fuel.

Rocketry

Rockets were invented in China more than 800 years ago. The first rockets were fueled by gunpowder, just like modern fireworks, and were used as weapons. The rocket was revived as a weapon of war in the late 1700s. Rockets were used by the British against the Americans in the War of 1812. The event is commemorated by the words "rockets' red glare," in the American national anthem.

By the end of the 1800s, scientists in many countries were considering rockets as a means of propulsion. In 1895 the Russian scientist, K. E. Tsiolkovski, was the first to propose rockets as a means of space travel. He also foresaw the use of liquid rocket fuels, and invented the important concept of the multistage rocket. For many years, however, Tsiolkovski's work was given little recognition.

In the early 1900s, rocket research was taken up by the American scientist and inventor, Robert H. Goddard. In 1926, Goddard successfully launched the world's first liquid-fueled rocket. Although it traveled less than 60 m (200 ft.), it was an important first step.

During World War 2, German scientists turned the liquid-fueled rocket into a long-range weapon. In 1944, the V-2 rocket was first launched. It was able to carry a 1 metric ton explosive warhead for more than 300 km (up to 200 mi.). Later on, much bigger rockets were developed independently by both the United States and Russia.

By the early 1960s, both countries had developed large, multistage rockets. These rockets could be used to lift astronauts and artificial satellites into space.

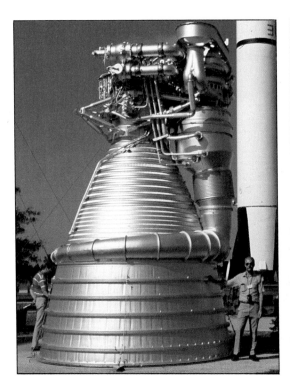

▲ One of the huge main engines fitted to the first stage of the Saturn V rocket. The large bell-shaped structure is the combustion chamber. Liquid fuel and liquid oxygen are supplied by pipes, valves, and pumps mounted above the combustion chamber.

▶ The American scientist Robert H. Goddard standing alongside the first liquid-fueled rocket in March 1926.

▶ Over the years, rockets have become progressively larger and have carried heavier payloads over greater distances. Goddard's experimental rocket of 1926 carried nothing but its own weight. The diagram (1) shows the rocket in the photo, drawn to scale. The German V-2 (2) carried a payload of high explosives. It was used during World War 2 to bombard cities hundreds of miles away. In 1961, the Vostok rocket (3) was used to launch the first man into space, the Russian Yuri Gagarin. Starting in 1967, the giant Saturn V (4) was used to launch a series of American Apollo missions. In 1969 a Saturn V launched *Apollo 11*, which carried the first astronauts to land on the Moon. All these rockets are drawn to scale.

Multistage rockets

No single rocket is powerful enough to lift itself and a payload into space. A number of rocket units, or stages, are needed. They are joined end to end to form a multistage rocket. The one illustrated has three stages and twin boosters for extra thrust at lift-off. The various stages fire in turn and then fall away, thrusting the satellite payload until it is traveling fast enough to go into orbit.

Liquids and solids

Steam engines, gasoline engines, and jets are all air-breathing engines. They take the oxygen required for combustion from Earth's atmosphere, and need only to be supplied with fuel. Rocket engines, however, carry their own supply of oxygen as well as their fuel, which gives them two important advantages. The performance of a rocket engine is not limited by the amount of air that it can take in. This means that rockets can produce far more thrust than jet engines of the same size. More importantly, carrying their own oxygen means that rockets can operate outside the atmosphere, in the near-vacuum conditions of space.

There are two main types of rocket: those that use solid fuel, and those that use liquid fuel. Solid-fueled rockets, of which gunpowder rockets are the commonest example, carry fuel and oxygen mixed in solid chemical form. Once the fuel has been ignited, it burns steadily until it has all been consumed. The advantage of solid-fueled rockets is that once manufactured, they are ready for immediate use. They are mainly used in missiles, and as launch boosters for bigger rockets.

Liquid fuels

Liquid-fueled rockets carry both fuel, such as hydrogen, and oxygen in liquid form, inside separate refrigerated tanks. Fuel and oxygen are injected into the combustion chamber at high pressure and are then ignited.

The main advantage of liquid-fueled rockets is that they can be shut down during flight, and then reignited when required. The power output of the engine can also be adjusted during flight by regulating the amount of fuel and oxygen passing through the pumps and injector. The main disadvantage of such rockets is that fuel and oxygen cannot be stored on board. The fueling operation, which can take several hours, must be carried out shortly before the rocket is to be launched. Liquid-fueled rockets are used in the largest intercontinental missiles and for space exploration.

▼ The complex flow of fuel and oxygen through a typical liquid-fueled rocket. Temperatures inside the combustion chamber can reach over 5,000°C (9,000°F), and some of the liquid hydrogen is used to cool the combustion chamber.

▲ Building one of the space shuttle's booster rockets inside the Vehicle Assembly Building. The exhaust nozzles are visible at the bottom. Most of the booster's mass consists of a tall column of solid fuel.

▲ Inside the Vehicle Assembly Building at the Kennedy Space Center in Florida, the space shuttle orbiter *Discovery* is lowered into position on its external fuel tank and solid-fueled rocket boosters. The external tank provides propellants for the orbiter's three main liquid-fueled engines.

▶ A battery of Nike-Hercules missiles. The Nike-Hercules missile has a solid-fueled rocket engine and is designed to shoot down aircraft at altitudes up to 30 km (nearly 20 mi.) above the Earth's surface. Liquid-fueled rockets could not be used for this purpose. They take too long to fuel and prepare for launch to be used in an emergency.

Rocket workhorses

The exploration of space has been dominated by Russia and the United States. More than 98 percent of the rockets launched into space have belonged to these two countries.

The main workhorse of the Russian space program has been the A-2 rocket. This is still used to launch the manned *Soyuz* and unmanned *Progress* spacecraft. Standing 49 m (161 ft.) tall on the launch pad, the A-2 has a main core stage and four boosters: all burn kerosene and liquid oxygen. More powerful Russian rockets include the Proton (D-1), and *Energiya* rockets, both about 60 m (200 ft.) tall. *Energiya* can lift 100 metric tons into orbit. It is also used to launch the Russian space shuttle.

▶ An American Delta rocket streaks away from the launch pad at Cape Canaveral, Florida. Its nine solid-fueled boosters provide most of the lift-off thrust.

▼ *Ariane 3*, developed and built by the European Space Agency, on the launch pad in French Guiana. The *Ariane* rockets were developed to launch satellites for countries without their own rockets.

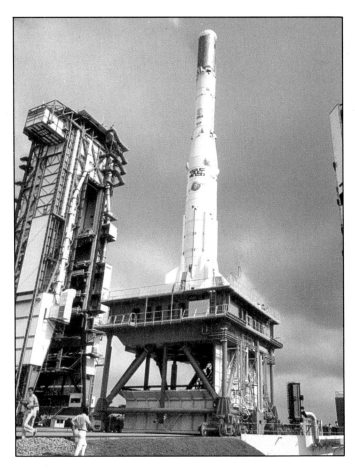

During the 1980s, the United States began to phase out its expendable rockets, such as the Delta and the Titan. It began to switch most of its satellite launchings to the reusable space shuttle instead.

The space shuttle hardware is made of three main units. A winged orbiter carries the crew and payload (cargo). An external tank carries fuel for the orbiter's three liquid-fueled engines. Two solid-fueled boosters, attached to the external tank, provide additional thrust at lift-off. The space shuttle can carry up to 20 metric tons of payload into low Earth orbit. After a mission the orbiter descends and lands on a runway like a normal aircraft.

▼ A Russian rocket being delivered to the launch pad at the Baikonur Cosmodrome in Central Asia. Hydraulic rams are used to lift it into the vertical position alongside the launch gantry. The Russians have more experience at launching rockets than any other country. In the former Soviet Union, rockets were set up and prepared for launch in about two days, compared with about 100 days in the United States.

The space shuttle has not proved as cheap or as reliable as hoped. Following the *Challenger* disaster in 1986, the United States started using expendable rockets again for routine satellite launchings. The shuttle is now used to support scientific missions. For example, it has been used to carry the space laboratory *Spacelab* and to launch probes to the planets.

The European Space Agency (ESA) has also developed a range of powerful launching rockets, called *Ariane*. They launch satellites for international customers on a commercial basis. *Ariane 3* is a three-stage launch vehicle that can place up to three satellites into geostationary orbit, 35,900 km (22,300 mi.) high. ESA launches the *Arianes* from its launch site at Kourou in French Guiana.

In the Far East, Japan, China, and India have also developed launch rockets. Japan has a thriving space program, and now launches space probes as well as satellites. In 1986 two of its probes encountered Halley's comet and returned much new information.

Units of measurement

Units of measurement

This encyclopedia gives measurements in metric units, which are commonly used in science. Approximate equivalents in traditional American units, sometimes called U.S. customary units, are also given in the text, in parentheses.

Some common metric and U.S. units

Here are some equivalents, accurate to parts per million. For many practical purposes rougher equivalents may be adequate, especially when the quantity being converted from one system to the other is known with an accuracy of just one or two digits. Equivalents marked with an asterisk (*) are exact.

Volume
1 cubic centimeter = 0.0610237 cubic inch
1 cubic meter = 35.3147 cubic feet
1 cubic meter = 1.30795 cubic yards
1 cubic kilometer = 0.239913 cubic mile

1 cubic inch = 16.3871 cubic centimeters
1 cubic foot = 0.0283168 cubic meter
1 cubic yard = 0.764555 cubic meter

Liquid measure
1 milliliter = 0.0338140 fluidounce
1 liter = 1.05669 quarts

1 fluidounce = 29.5735 milliliters
1 quart = 0.946353 liter

Mass and weight
1 gram = 0.0352740 ounce
1 kilogram = 2.20462 pounds
1 metric ton (tonne) = 1.10231 tons

1 ounce = 28.3495 grams
1 pound = 0.453592 kilogram
1 short ton = 0.907185 metric ton (tonne)

Length
1 millimeter = 0.0393701 inch
1 centimeter = 0.393701 inch
1 meter = 3.28084 feet
1 meter = 1.09361 yards
1 kilometer = 0.621371 mile

1 inch = 2.54* centimeters
1 foot = 0.3048* meter
1 yard = 0.9144* meter
1 mile = 1.60934 kilometers

Area
1 square centimeter = 0.155000 square inch
1 square meter = 10.7639 square feet
1 square meter = 1.19599 square yards
1 square kilometer = 0.386102 square mile

1 square inch = 6.4516* square centimeters
1 square foot = 0.0929030 square meter
1 square yard = 0.836127 square meter
1 square mile = 2.58999 square kilometers

1 hectare = 2.47105 acres
1 acre = 0.404686 hectare

Temperature conversions

To convert temperatures in degrees Celsius to temperatures in degrees Fahrenheit, or vice versa, use these formulas:

Celsius Temperature = (Fahrenheit Temperature − 32) × 5/9
Fahrenheit Temperature = (Celsius Temperature × 9/5) + 32

Numbers and abbreviations

Numbers

Scientific measurements sometimes involve extremely large numbers. Scientists often express large numbers in a concise "exponential" form using powers of 10. The number one billion, or 1,000,000,000, if written in this form, would be 10^9; three billion, or 3,000,000,000, would be 3×10^9. The "exponent" 9 tells you that there are nine zeros following the 3. More complicated numbers can be written in this way by using decimals; for example, 3.756×10^9 is the same as 3,756,000,000.

Very small numbers – numbers close to zero – can be written in exponential form with a minus sign on the exponent. For example, one-billionth, which is 1/1,000,000,000 or 0.000000001, would be 10^{-9}. Here, the 9 in the exponent -9 tells you that, in the decimal form of the number, the 1 is in the ninth place to the right of the decimal point. Three-billionths, or 3/1,000,000,000, would be 3×10^{-9}; accordingly, 3.756×10^{-9} would mean 0.000000003756 (or 3.756/1,000,000,000).

Here are the American names of some powers of ten, and how they are written in numerals:

1 million (10^6)	1,000,000
1 billion (10^9)	1,000,000,000
1 trillion (10^{12})	1,000,000,000,000
1 quadrillion (10^{15})	1,000,000,000,000,000
1 quintillion (10^{18})	1,000,000,000,000,000,000
1 sextillion (10^{21})	1,000,000,000,000,000,000,000
1 septillion (10^{24})	1,000,000,000,000,000,000,000,000

Principal abbreviations used in the encyclopedia

°C	degrees Celsius
cc	cubic centimeter
cm	centimeter
cu.	cubic
d	days
°F	degrees Fahrenheit
fl. oz.	fluidounce
fps	feet per second
ft.	foot
g	gram
h	hour
Hz	hertz
in.	inch
K	kelvin (degree temperature)
kg	kilogram
l	liter
lb.	pound
m	meter
mi.	mile
ml	milliliter
mm	millimeter
mph	miles per hour
mps	miles per second
mya	millions of years ago
N	north
oz.	ounce
qt.	quart
s	second
S	south
sq.	square
V	volt
y	year
yd.	yard

Glossary

afterburning Technique used to increase the power of jet engines, in which fuel is also burned in a second combustion chamber located behind the turbine.

alternating current Electricity supplied to houses and factories, in which the current reverses direction, usually many times per second.

anthracite The best-quality coal. A hard, black substance containing at least 86 percent solid carbon. Anthracite gives off little smoke when burned, and is mainly used in household heating.

anticline Dome-shaped formation of underground rock strata in which oil and natural gas often accumulate. About 80 percent of world oil production comes from anticlines.

atom Smallest individual unit of an element. Consists of a nucleus surrounded by one or more orbiting electrons.

beam engine Type of steam engine in which the up-and-down motion of the piston is transferred to a hinged lever or beam.

bimetallic Constructed of two different metals that expand and contract at different rates when heated or cooled.

bituminous coal The most widely distributed form of coal, consisting mainly of solid carbon, but also containing significant quantities of coal tar and gas.

boosters Solid-fueled rocket motors that are often used to provide additional thrust for launching or take-off. Boosters are normally discarded after their fuel has been exhausted.

branch circuit Type of electrical circuit used to distribute electricity to the different parts of a room or floor of a house.

brown coal Low-grade coal that is often soft enough to be crumbled between the fingers, and which is mined in some countries for use in power plants.

bus system Network of thick copper wires and bars used in power stations and substations to carry high-voltage electricity.

camshaft Metal shaft with a series of rounded projections that is used to operate the valves in a four-stroke engine.

carbonization Process by which dead plant material turns to carbon. The term is usually applied to the formation of coal through the action of heat and pressure within the Earth's crust.

carburetor Mechanical device for mixing air and gasoline in the correct proportions for combustion in a gasoline engine.

catalyst Any substance that enables a chemical reaction to take place, or which speeds up a chemical reaction, but which is not itself changed by the reaction.

catalytic converter Device that can be installed in the exhaust system of a motor vehicle to reduce the amount of pollution given off in exhaust gases.

cavity magnetron Device that converts electricity into microwaves, and is used in radar systems and in microwave ovens.

chain reaction Continuous process in which atoms of, for example, uranium split into smaller atoms. It is kept going by the action of neutrons.

charcoal Black, crumbly substance, consisting almost entirely of carbon, that is made by burning wood in the absence of oxygen.

Christmas tree System of pipes and valves at the top of many oil wells. It is used to regulate the pressure of the oil reaching the surface.

circuit breaker Safety device that breaks an electrical circuit if it becomes overloaded or if there is a sudden power surge.

coal face The part of a coal seam that is being worked in an underground mine.

coal tar Thick, sticky liquid that can be extracted from coal. It contains many useful chemicals.

coke Coal that has been baked in an oven to remove any gas and coal tar. Coke gives off little smoke when burned and is mainly used by industry and power plants.

combustion chamber That part of a jet or rocket engine in which fuel is burned.

commutator Part of an electric motor or generator (direct current only) that passes current in to or out of the coil by rotating against contacts known as brushes.

compound turbine Type of steam turbine that uses steam at different pressures in a series of cylinders.

condenser Part of a steam engine or steam turbine in which steam is condensed after it has been used to produce power.

containment Thick layer of reinforced concrete that surrounds a nuclear reactor as a safety precaution to prevent the escape of harmful radioactivity.

core Central part of a nuclear reactor in which a chain reaction takes place. Consists typically of uranium fuel rods, a moderator, and coolant.

cracking Process used in the refining of crude oil, in which large hydrocarbon molecules are broken down into smaller ones.

crankshaft That part of a gasoline or diesel engine that transforms the up-and-down movement of the pistons into rotary motion.

Darreius turbine Type of vertical-axis wind turbine with curved blades that are attached to the rotor at each end.

decomposition Process by which dead plant and animal material is broken down by the action of bacteria and other microorganisms.

direct current Electricity in which the current flows in one direction only. Its main use is in battery-powered devices (batteries can produce only direct current), but it is also used in some industrial processes.

discharge lamp Type of electric lighting that produces light by exciting the atoms of a gas or vapor. A fluorescent lamp is a special kind of discharge lamp. Ordinary discharge lamps, however, are mainly used for street lighting and advertising because they produce only light of one particular color.

distributor Part of the electrical system of a gasoline engine that supplies high-voltage current to each spark plug in turn.

double-acting Term used to describe steam engines in which every stroke of the piston is driven by steam.

enrichment Process that increases the ability of uranium to sustain a chain reaction, and which therefore makes it a more useful and valuable fuel.

fast breeder reactor Type of reactor that does not require a moderator, and which can be used to create plutonium from uranium fuel.

filament Coil of very fine wire found inside electric light bulbs. Light is produced by the filament glowing white hot.

fission Process by which a large atom splits into two smaller atoms, releasing energy.

flat-plate collector Simple device for collecting solar energy that consists of a water-filled coil behind a glass cover. Flat-plate collectors are normally used for domestic hot water and are usually installed on rooftops.

flywheel Heavy wheel that smooths out the rotary motion produced by reciprocating engines.

fossil Any physical remains of ancient life. Most fossils are just impressions in stone; only under special circumstances is any organic material preserved.

four-stroke engine Type of engine in which power is produced by only one piston stroke in every four.

fractional distillation Basic process of oil refining by which the different hydrocarbons are separated out at different temperatures inside a hollow metal column.

fuel injector Device that injects controlled amounts of vaporized fuel into the cylinder of a four-stroke engine. Fuel injectors are found on all diesel engines, but on only some gasoline engines.

fuel rods Long metal cylinders into which uranium fuel is placed before it is loaded into a nuclear reactor.

fuse Safety device in an electrical circuit. Usually a short piece of wire that melts if the circuit becomes overloaded.

gallery Side tunnel in an underground mine, normally dug at right angles to the main tunnel.

gas turbine A turbine engine powered by hot, compressed gas. The name is sometimes applied to jet engines with turbines.

generator Device used to produce electric current by rotating a coil of wire in a magnetic field.

geothermal Relating to heat energy within the Earth's crust caused by natural radioactivity.

graphite Naturally occurring form of pure carbon.

gravity The downward pulling force of the Earth and other planets and stars. The force of attraction between two objects.

head (of water) Vertical distance traveled by running or falling water.

heat exchanger Device used to transfer heat energy from one fluid to another, but without them coming into direct physical contact.

heating element Coil of resistant wire that gives off useful quantities of heat when an electric current is passed through it.

hydrocarbons Large group of chemical compounds, the molecules of which consist almost entirely of hydrogen and carbon atoms. Crude oil contains thousands of different hydrocarbons and is the main source of these substances.

hydroelectric power Electricity that has been produced by the action of moving water against the blades of a turbine.

ignition coil Electrical device that produces the high voltages needed by the spark plugs of a gasoline engine.

impermeable Waterproof or in some other way impenetrable to a liquid or gas. The term is normally applied to rocks that do not permit water to pass through them.

impulse turbine Form of turbine that is rotated by the impact of a fluid, for example steam or water, against a series of cups or buckets mounted around the rim of a wheel.

insulation Term applied to any way of reducing the unwanted transfer of energy. Heat insulation keeps things hot or cold; electrical insulation confines the flow of electricity.

insulator Any substance that is an extremely bad conductor of electricity, for example glass and ceramics. The term also refers to objects made of these substances that are used to insulate electrical wires or cables.

internal-combustion engine Any engine that burns fuel inside the engine itself in order to produce power.

jet engine Any engine that burns a mixture of fuel and compressed air to provide a stream of hot exhaust gas, which propels the engine forward.

kerogen Partially decomposed plant and animal material from which oil is formed by heat and pressure within the Earth's crust.

lignite Lowest grade of coal; a brown crumbly substance mainly burned in power plants.

liquefied petroleum gas Collective name for hydrocarbons such as butane and pentane, which are gases at normal temperature and pressure, but which are easily liquefied for storage.

longwall Method of underground mining in which coal is taken from a seam along the entire side of a gallery. As the coalface moves back, the roof of the gallery is allowed to collapse.

mantle The stony section of the Earth between the crust and the core.

methane The lightest hydrocarbon and the main constituent of natural gas.

microwaves High-frequency radio waves that can induce heat in objects that they pass through.

migration (of oil and gas) Process by which oil and gas tend to move from one location to another within the Earth's crust.

missile Any nonliving object that travels through the atmosphere before striking the ground or another object. The term is usually applied to rockets used in warfare.

moderator Substance used in the core of a nuclear reactor to slow down, or moderate, neutrons in order that a chain reaction can take place.

molecule Smallest unit of a particular substance consisting of a characteristic arrangement of atoms.

neutron One of the two types of subatomic particle (the other being a proton) that make up the nucleus of most atoms.

nuclear Relating to an atomic nucleus. The term is often used to refer to the production of energy by means of a chain reaction.

nucleus Central part of an atom, in most cases consisting of both protons and neutrons.

oil pan Main lubricating-oil reservoir of a four-stroke engine.

oil shale Type of oil deposit in which crude oil is trapped within a fine-grained rock, and can be removed only with great difficulty.

overburden Rocks and soil covering a coal seam that lies just below the surface.

payload Carrying capacity of a rocket or missile, given in terms of weight.

peat Waterlogged and partially decomposed plant material that is the first stage in the formation of coal. In some parts of the world, dried peat is used as a fuel.

penstocks Water inlets in the face of a dam that channel water to the turbines.

permeable Describes any substance that allows water or some other liquid or gas to pass through it. The term is usually applied to rock.

photovoltaic cell Device that produces small amounts of electrical current when exposed to sunlight. Often called a solar cell.

plutonium Artificial element that does not occur naturally, but is produced inside nuclear reactors.

pollution Unnatural presence of any substance in the environment, usually with harmful effects.

power tower Type of central collection system for producing electricity from solar energy. A large number of mirrors reflect the Sun's rays to the top of the tower, producing enough heat to boil water for turbines.

pumped storage Method of indirectly storing electricity. Surplus electrical power is used to pump water uphill to a reservoir. When additional electricity is required, the water is allowed to flow downhill to drive turbines.

pylon Tall metal framework used to support overhead power lines.

radioactivity Changes in nuclei of certain substances, including uranium and plutonium, which cause the emission of harmful radiation.

ramjet Simplest form of jet engine. It has no compressor or turbine.

reaction engine Any engine that produces power through the thrust of the exhaust gases, for example jets and rockets.

reaction turbine Type of turbine that is driven by the flow of a fluid through a series of angled blades attached to a shaft. Most turbines used today, whether driven by water, steam, or gas, are reaction turbines.

reactor Basic component of a nuclear power plant. A device for producing heat from a slow, steady chain reaction. A reactor consists of a core, a containment vessel, and a cooling system.

reciprocating engine Any engine that produces power through the action of a piston within a cylinder, for example steam, gasoline, and diesel engines.

rocket 1. Type of engine that burns a mixture of fuel and oxygen which it carries with it, and which therefore operates outside the Earth's atmosphere. **2.** Any vehicle propeled by a rocket engine; usually applied to vehicles designed to launch objects or people into space.

room-and-pillar Method of underground mining in which coal is removed from a seam in a series of square "rooms," with pillars of coal being left in place to support the roof.

rotary engine Type of engine that uses a specially shaped "piston" to provide direct rotary motion rather than an up-and-down movement.

rotor That part of a mechanical device, for example a turbine or an electric motor, that produces rotary motion.

runner Alternative name for the rotor in some types of water turbine.

sedimentary rock Rock formed from layers of sediment laid down in rivers, lakes, and seas.

slip rings Parts of an electric motor or generator (alternating current only) that pass current in or out of the coil.

solar gain Heat energy that buildings receive from sunlight falling on walls, roofs, and through windows.

staging Method of improving the efficiency of steam turbines by allowing the steam to lose its energy in a series of stages.

strip mining Type of surface mining. The coal may be removed from a single large hole that has been dug down to the seam. Alternatively, it may be removed from a series of trenches. The overburden from one trench is used to fill in the previous trench.

substation Installation that reduces the voltage of electrical current by passing it through transformers, before it is distributed.

superheated Describes water or steam that has been heated under pressure to above 100°C (212°F).

switchgear System of electrical switches found in power stations and substations.

traction engine Steam-powered tractor used especially during the late 1800s.

transformer Device used to increase or reduce the voltage of alternating current.

Trombe wall Simple solar heating device that can be built into houses and other buildings. A double wall, with an outer wall of glass separated from a solid inner wall by air.

turbine Any device that produces rotary motion through the action of a gas or liquid on angled blades attached to a shaft.

turboalternators Electrical generators, driven by turbines, which produce alternating current.

turbofan Type of jet engine fitted to aircraft that use the turbine to power a fan at the front of the engine. This fan produces most of the engine's thrust.

turbojet Basic jet engine that uses a turbine driven by the exhaust gases to compress air for combustion with kerosene fuel. All the thrust is produced by the exhaust gases.

turboprop Type of jet engine that uses the turbine to power a propeller mounted ahead of the engine. Both the exhaust gases and the propeller produce thrust.

two-stroke engine Simple gasoline engine that produces power with every other stroke of the piston, mainly used in lawnmowers and small motorcycles.

uranium Heaviest naturally occurring element and a principal natural source of nuclear fuel.

volute chamber Spiral casing surrounding the rotor of most types of water turbine.

wellheading Process by which oil can be loaded onto a tanker directly from an undersea oil well.

wind turbine A windmill that is designed to generate electricity.

Index

Page numbers in *italics* refer to pictures. Users of this Index should note that explanations of many scientific terms can be found in the Glossary.

A

absorption 24
acids 24
aeolipile 51, *51*
afterburning 79
aircraft 76, 77, *79*
Alaska 16, 17
Algeria 23
alternating current 49, 56, 57
anthracite 10, 12, 13
anticline 18, *19*
Apollo 14 81
Arctic Circle 23
Argentina 52
argon 62
Ariane 3 76, 84, *84*, 85
asphalt 18, 24
atom, structure of *28*
atomic bomb 28
Australia 8, 9, 13, 32, 36, 40, 47, 74
aviation fuel 16

B

Baikonur Cosmodrome 85
barrels 22
beam engine 67
benzene 18
bitumen 18
bituminous coal 10
Boeing 707 77
boosters 86
branch circuit 64
Brazil 20, 48, 75
breeder reactor 31
brown coal 8, 10, 12, 13
bucket-wheel excavator 8, *13*
bus room 56
buses 74
butane 19, 22, 24

C

calculator, solar-powered 32
camshaft *71*
Canada 52
CANDU reactor 52, *52-53*
cap rock 18

carbon 9, 17
carbon black 24
carbon dioxide 26, 30
carbonization 10
catalysts 24
catalytic converter 72
cavity magnetron 63
central heating 65
chain reaction 28, 29, 30, 52
charcoal 9
chemical industry 59
Chernobyl 27, 52
China 9, 16, 20, 38, 44, 80
circuit breaker 56, 64
clays 24
coal 8-15
   carbon in 10
   formation of 10, *11*
   grading of 15
   origins of 8
   reserves of 9, *9*
   seams 9, 12
   transport of 15
   types of 10
coal face 14
coal tar 9, 10
coalfield 8, 9
coil 62
coke 10
communications 59
commutator 60
compound turbine 51
computing 59
condensers 48
condensing plates 24, *24*
containment 30
control rods 30, 52
conveyor belts 15, *15*
coolant 30
coolant system 52, 53, 56
cooling system, car *72*, 73
cracking 24
crude oil 16, 18, 24
cylinder 70, 72

D

Daimler, Gottlieb 70
dams 41, *41*, 45
Darreius wind turbines *42*, 54
De Havilland Comet 77
decomposition 10, 11, 18
Delta rocket 84, *84,* 85
Denmark 38
diesel engines 66, 74
diesel fuel 16, 24
Diesel, Rudolf 74
dinosaurs 8

direct current 49, 57
distillation 24
   fractional 24, *24*
distributor 73
draglines 12
drilling 20
   angled 20
   offshore 20, 21, *21*

E

earthquakes 16, 22
Edison, Thomas A. 49, 59
electric circuits 64
electric lighting 58, 62, *63*
electric space heater 59
electric wiring 64
electricity 9, 27, 32, 38, 39, 47, 48, 58
   distribution of 60, 61
   generation of 49
   from solar energy 36, 37
   uses of 59, 62, 63
electromagnet 49, 61, 65
electrons 28, *28*
Energiya rocket 76, 84
energy 38, 47, 48
engine systems 72
engines 47, 66-75
   air-cooled 73
   diesel *see* diesel engines
   gasoline *see* gasoline engines
   rotary *see* rotary engines
   steam *see* steam engines
environment 12, 29, 48, 52
ethane 19
Europe 8, 11, 14, 27, 49, 69
European Space Agency (ESA) 84
exhaust system 72
expansion chamber 68
explosions 20
external-combustion engine 66

F

factories 60
Faraday, Michael 49
farming 66, 69, 74
farmland 12
fast-breeder reactor 31
fault *19*
feedstock 24
fission 28, 29
flat-plate collector 34
floodlighting 62
fluorescent light 62, *62*
fossil fuel 56

92

four-stroke engine 70, *70-71*
fractionating column 24, *24*
fractions 24
France 26, 27, 29, 38, 44, 60
Francis, James 54
Francis turbine 54, *54*
fuel injector 72, 72
fuel rods 28, *28,* 30
fuel system 72
fuels 17, 24, 70, 75

G

Gargarin, Yuri 81
gas, natural 16, 17
   deposits 19
   formation of 19
   liquid 22
gas separation plant 24
gas-turbine engine 77
gasoline 24
gasoline engine 66, 70, 71, *71*, 73
generator 49, 52, *53*, 56, *57*, 73
geothermal energy 38, 44
Germany 12, 49, 70, 74, 77
Geysers power plant 4, *44*
Goddard, Robert H. 80, *80*
graphite 30
Great Britain 14, 34
grid 60, *60-61*
gunpowder 80

H

hairdryer *65*
hard coal 10, 13
heat exchanger 30, 52, 53
heating element 62
heating fuel 16, 17, 24
heavy oil 20, 24
Hero 51
hexane 18
household appliances 59
Hungary 44
hydrocarbons 17, 18, *18*, 24
hydroelectric power plants 38, 40, *40*, 48
hydrogen 17, 24, 32, 76

I

ignition 70
impulse turbine 50
incongruity *19*
India 52
Industrial Revolution 9, 10, 39

information technology 59
infrared light 62
insulation 34
insulator 60
intercontinental missile 82
internal-combustion engine 66, 70
intrusion *19*
Italy 36, 38, 44

J

Japan 36, 44
jet engine 76, 77, 78-79, *78*
jet plane 76
joulemeter 64

K

Kaplan turbine 54, *54*
Kennedy Space Center 83
kerogen 18
kerosene 24
kinetic energy 50

L

light bulb 58, 59, *59*, 62
   invention of 59
   principles of 62
lignite 10, 12, 13
limestone 18
liquefied natural gas (LNG) 22
liquid fuel, in rockets 82
   refrigeration of 82
liquid methane 22
locomotives 66
   diesel 74
longwall mining 14
LPG tanker *22*
lubrication system 72, 73
   in 2-stroke engines 75

M

mechanical excavator 12, *13*
Mediterranean Sea 23, 34
mercury 62
methane 16, 18, 19, 24
microwave oven 63, *63*
mines 12-15, *15*
mining 12-15
   longwall 14
   room-and-pillar 14
   strip 12

   surface 12, 13
   underground 14, 15
moderator 29, 30
molecular sieves 24
motor 49, 58, 65, *65*
motorcycle 73, *74*

N

neon 62
neutrons 28, 30
New Zealand 44
Newcomen, Thomas 67
nickel-chromium alloy 59
Nike-Hercules missile *83*
North Sea 16, 20, 23
nuclear energy 27, 28
nuclear fuel 29, 52, 56
nuclear power 26-31
nuclear power plants 26, 28, 38, 52, *52-53*
   safety systems 53
nuclear reactor 28, 30, 31

O

offshore oil platform 21
oil 16-25
   crude 16, 18
   formation of 18
   migration of 18
   production of 20
   reserves of 17, *17*
   transport of 22
oil, engine 72
oil field 16, 17
oil pipeline 16
oil pump *73*
oil refinery 24
oil reservoir 72
oil shales 20
oil traps *19*
Otto, N. A. 70
overburden 12
oxygen 11, 82

P

Paraguay 48
Parsons, Charles 49
payload, rocket 81, 85
Pearl Street power station 49
peat 10, 11, *11*
Pelton wheel 54, *55*
pentane 19
Persian Gulf 22

petrochemicals 24
petroleum 16
phosphors 62
photovoltaic cell 36
pinch-out *19*
pipelines 22, *23*, 31
plutonium 26, 31
pollution 8, 27
power 47
power circuit 64
power lines 60
power plants 8, 15, 48-49, 51, 56-57
power shovels 12
pressurized-water reactor 30, *30*, 31
propane 19, 22
Proton rocket 84
protons 28
pumped-storage hydro plant 40
pylons 60

R

radar 63
radiation 27
radioactive waste 26, *29*
radioactivity 26, 27, 29, 30, 52
railroads 69, 74
Rance River 45
reaction engine 76
reaction turbine 50, 54
reactor vessel 30
reciprocation engine 66
refineries 24
reservoirs
   oil and gas 17
   water 40
resistance 62
road train *74*
rockets 76, 80-85
   liquid-fueled 80, 82, *82*
   multistage 80, 84
   solid-fueled 82
rotary engine 74
rotary power 49, 67, 70, 77
rotor 50
Russia 9, 44, 76, 84

S

salt dome 19
Salter duck 45, *45*
sandstone 18
satellites 84
Saturn V *80, 81*
Savery, Thomas 67
seams, coal 9, 12, 14

sediment 10, 18
shafts 14, *15*
Siberia 16, 17, 23
silicon 36
smoke 10, 11
sodium 31, 36, 62
soft coal 10
solar cells 33, 36, *36*
solar energy 32, 33, *33*, 34-37
solar field 36
solar furnace 36
solar gain 34, *34, 35*
solar heating 32, 33, 34
solar-energy power station 32
solar-powered car 33
solvents 24
Soviet Union (former) 20, 80
Soyuz rocket 84
space shuttle 76, *83*, 85
Spain 36
speed of sound 76
steam engine 66, 67, 68-69
steam pump 67
steam turbines 30, 48-49, 50-51, *50, 51*, 69
   staging in 51
substations 60
Sun 32, 33
Sun tracking 36
surveying 20
Swan, Joseph 59
switch, bimetallic 65
switchgear 56

T

tankers 21, 22
television 59
thermostat 72
Three Mile Island 27, 31, 52
tidal power 38, 45
traction engine 66, *68*, 69
transformers 56, 57, 60, *61*
Trombe wall 35, *35*
Tsiolkovski, K. E. 80
turbines 48, 52
   compound *see* compound turbine
   horizontal-axis 54
   impulse *see* impulse turbine
   Kaplan *see* Kaplan turbine
   reaction *see* reaction turbine
   steam *see* steam turbine
   vertical-axis 54
   water *see* water turbines
   wind *see* wind turbines
turboalternator 56
turbofan 77, *77*, 79, *79*
turbogenerator 56

turbojet 77, *78*
turboprop 77, 79, *79*
two-stroke engine 74, 75, *75*

U

Ukraine 26, 27
United States 8, 9, 12, 13, 14, 16, 23, 27, 32, 36, 38, 44, 48, 49, 56, 76, 80, 84, 85
uranium 26, 28, 29
uranium fuel 28, *28*, 30

V

V-2 rocket 80, *81*
vacuum pump 67
vacuum-tube collector 35
VLCCs 22
volcanoes 19, 44
voltages 56, *57*, 59, 61
volute chamber 54
Vostok *81*

W

Wankel engine 74
water cycle 38
water pump 65
water turbine 54, *54*, 55
water-cooled reactor *30*, 31
waterpower 38
waterwheel 39
Watt, James 67
wave power 38, 40, 41
wellheading 22
wet gas 19
Whittle, Frank 77, *77*
wind power 38, 42
wind turbine 38, 42, 43, 48, 54
winding gear 14, *15*
windmill 54

X

X-1 76

Y

Yeager, Charles "Chuck" 76

Further reading

Andryszewski, Tricia. *What to Do about Nuclear Waste*. Brookfield, CT: Millbrook Press, 1995.
Bardon, Keith. *Exploring Forces and Structures*. Chatham, NJ: Raintree Steck-Vaughn, 1992.
Catherall, Ed. *Exploring Uses of Energy*. Chatham, NJ: Raintree Steck-Vaughn, 1990.
Charman, Andrew. *Energy*. Danbury, CT: Franklin Watts, 1993.
Dineen, Jacqueline. *Oil, Gas and Coal*. Texas: Raintree Steck-Vaughn, 1995.
Friedhoffer, Robert. *Forces, Motion, and Energy*. Danbury, CT: Franklin Watts, 1992.
Friedhoffer, Robert. *Matter and Energy*. Danbury, CT: Franklin Watts, 1992.
Galperin, Anne L. *Nuclear Energy – Nuclear Waste*. New York: Chelsea House, 1992.
Gardiner, Brian. *Energy Demands*. New York: Gloucester Press/Watts, 1990.
Gardner, Robert. *Forces and Machines*. Morristown, NJ: Silver Burdett Press, 1991.
Hawkes, Nigel. *Nuclear Power*. Vero Beach, FL: Rourke Corporation, 1990.
Hooper. *Electricity*. Texas: Raintree Steck-Vaughn, 1994.
Kerrod, Robin. *Force and Motion*. Tarrytown, NY: Marshall Cavendish, 1993.
Kerrod, Robin, *Future Energy and Resources*. New York: Franklin Watts, 1990.
Kerrod, Robin. *Natural Resources*. Tarrytown, NY: Marshall Cavendish, 1994.
Lampton, Christopher. *Nuclear Accident*. Brookfield, Connecticut: Millbrook Press, 1992.
Sauvain, Philip. *Motion*. Old Tappan, NJ: Simon & Schuster Children's, 1992.
Tayor, Barbara. *Force and Movement*. Danbury, CT: Franklin Watts, 1990.
Wellington, Jerry. *The Super Science Book of Energy*. New York: Thomson Learning, 1994.
Wellington, Jerry. *The Super Science Book of Forces*. New York: Thomson Learning, 1994.

Picture Credits

b=bottom, t=top, c=center, l=left, r=right.

ARPL Ann Ronan Picture Library, Somerset. FSP Frank Spooner Pictures, London. MARS Military Archive and Research Services, Braceborough. NASA National Aeronautics and Space Administration, Washington. RF Rex Features, London. RHPL Robert Harding Picture Library, London. SC Spacecharts, Wiltshire. SP Sci Pix,Wiltshire. SPL Science Photo Library, London.

6 SPL/Tom McHugh. 10 Spectrum Colour Library. 11b Zefa/G.Kait. 12 RHPL. 13t SPL/Tony Craddock. 13b Colorific/J.Howard. 14 Zefa. RHPL. 20l SPL/Sindo Farina. 20r Robin Kerrod. 22 SP. 23 Tony Stone Worldwide. 23 inset FSP. 25 Zefa. 27 RF. 27 inset Associated Press. 31 SP. 34 RF. 35 SPL/Tom McHugh. 36 Spacecharts.

37 FSP. 39l, 39r Zefa. 40 Central Electricity Generating Board. 42 SPL/Tom McHugh. 42-43 SPL/Lowell Georgia. 44 Robin Kerrod. 45 Zera/J. Pfaff. 46 SPL/Robin Scagell. 49l ARPL. 49r Michael Holford/Royal Institution. 50 Zefa. 52-53 Quadrant Picture Library/*Nuclear Engineering*. 55l G. R. Roberts. 55r Jerry Mason *New Scientist*. 56bl SPL/Hank Morgan. 56br SPL/U. S. Dept of Energy. 57t Picturepoint Ltd. 57b Zefa/J. Pfaff. 59l ARPL. 59r Michael Freeman. 61 Picturepoint Ltd. 63t Robin Kerrod. 63b Panasonic. 67l The Ridgeway Archive. 67r ARPL. 68 John Watney Photo Library. 70 All Sport/ Vandystadt. 71 Ford Motor Company. 74-75 Spectrum Colour Library/ D. J. Heaton. 75b, 77l Quadrant Picture Library. 77r SP. 78 Zefa/E. M. Bordis. 79t SC. 79cr SC/NASA. 80l Robin Kerrod. 80r NASA. 83tl, 83tr SC. 83b MARS/U. S. Army. 84l European Space Agency. 84r SC. 85 SPL/Novosti.